Studies in Theoretical Physics, Volume 2

Advanced mathematical methods

Online at: https://doi.org/10.1088/978-0-7503-4861-4

Studies in Theoretical Physics, Volume 2

Advanced mathematical methods

Daniel Erenso

*Department of Physics and Astronomy, Middle Tennessee State Univeristy,
Murfreesboro, Tennessee 37132, USA*

IOP Publishing, Bristol, UK

ISBN   978-0-7503-4861-4 (ebook)
ISBN   978-0-7503-4859-1 (print)
ISBN   978-0-7503-4862-1 (myPrint)
ISBN   978-0-7503-4860-7 (mobi)

DOI   10.1088/978-0-7503-4861-4

Version: 20240201

IOP ebooks

British Library Cataloguing-in-Publication Data: A catalogue record for this book is available from the British Library.

Published by IOP Publishing, wholly owned by The Institute of Physics, London

IOP Publishing, No.2 The Distillery, Glassfields, Avon Street, Bristol, BS2 0GR, UK

US Office: IOP Publishing, Inc., 190 North Independence Mall West, Suite 601, Philadelphia, PA 19106, USA

To my family: Your steadfast support, understanding, and encouragement have propelled my work to new heights. This dedication serves as a token of appreciation for the boundless inspiration, unwavering patience, and shared joy that have fueled the success of my past, present, and future endeavors.

Contents

Preface

This book is the second volume in the "Studies in Theoretical Physics" series, titled "Advanced Mathematical Methods." This comprehensive series aims to provide students majoring in physics or engineering with the essential mathematical tools required for advanced theoretical physics courses. This volume, which follows the foundational "Fundamental Mathematical Methods," addresses the critical role of Tensor calculus and its applications in courses such as General relativity, electromagnetism, fluid dynamics, quantum information theory, quantum field theory, solid mechanics, crystallography, continuum mechanics, plasma physics, and thermodynamics.

Diverging from standard textbooks in mathematical methods, this volume focuses exclusively on mathematical techniques vital for upper-level theoretical physics courses. The content is organized into four chapters, each building upon the vector calculus covered in the first volume. The first chapter introduces manifolds, and the second describes vector calculus on manifolds, drawing examples from the special theory of relativity. The third chapter explores tensor calculus on manifolds, while the fourth chapter applies tensor calculus, specifically to relativistic electrodynamics in the four-dimensional Minkowski space-time manifold.

Key features of this volume include its unique emphasis on mathematical techniques tailored for advanced physics and engineering courses, encompassing a diverse range of theoretical physics disciplines. The inclusion of numerous solved problems, complemented by modern 3D graphics, enhances the learning experience. Clear, brief, and concise discussions have been crafted to facilitate understanding for readers with English as a second language.

Acknowledgements

I am grateful to the physics major students at the Department of Physics & Astronomy at Middle Tennessee State University (MTSU) for their invaluable feedback and support that have influenced the content and direction of this book. My appreciation extends to the Department of Physics & Astronomy at MTSU for granting me the opportunity to teach upper-level physics courses for more than two decades. This book, along with its first volume, aspires to provide students with the essential mathematical methods required for advanced theoretical physics courses. I want to express profound thanks to my family and friends for their enduring tolerance and understanding as I dedicated countless hours to my past, present, and future books for more than two decades. Their consistent support has been my anchor, and I am truly thankful for their patience and encouragement. A special acknowledgment goes to Mr. Tesfu Kassaye Woldeyes, whose expertise and commitment have significantly elevated the quality of this book. Mr. Woldeyes played a pivotal role in editing the entire manuscript, and his insightful feedback has enhanced the clarity and coherence of the material.

Author biography

Daniel Erenso

Dr. Daniel Erenso is a Professor of Physics at Middle Tennessee State University (MTSU), Murfreesboro, Tennessee, USA. He joined MTSU in 2003 after he received his Ph.D. in theoretical physics from the University of Arkansas, a BSc (1990) and MSc (1997) in physics from Addis Ababa University in his native country Ethiopia. He also received an Advanced Diploma in Condensed Matter Physics from Abdul Salam International Center for Theoretical Physics (ASICTP), Trieste, Italy, in 1999. For more than two decades, Professor Erenso has served in teaching, research, and mentoring at different universities. In teaching, for the excellence and dedication that Dr. Erenso demonstrated, he received the MTSU College of Basic & Applied Sciences Excellence in Teaching Award in 2011. More recently, Professor Erenso has published two books: "Virtual and Real Labs for Introductory Physics II Optics, modern physics, and electromagnetism" and "Studies in Theoretical Physics, Volume 1: Fundamental Mathematical Methods". Professor Erenso has also maintained an active research program. He has been a research advisor for several undergraduate and graduate students. His research interests include theoretical and experimental physics and has published over 40 and presented over 80 research works at national and international venues. For his outstanding research accomplishment, Dr. Erenso received Sigma Xi the Scientific Research Society Aubrey E Harvey Outstanding Graduate Research Award from the University of Arkansas, the MTSU Foundations Special Project Award, MTSU, College of Basic & Applied Sciences Distinguished Research, and a nomination for American Physical Society (APS) Prize for a Faculty Member for Research in an Undergraduate Institution. Professor Erenso is a member of several professional societies that include the American Physical Society, Optical Society of America, American Association of Physics Teachers, Sigma Xi The scientific Research Society, and Sigma Pi sigma the Physics Honor Society. He has been serving as invited reviewer for international research journals and book publishing company since 2003.

Introduction

This second volume aims to equip students majoring in physics or engineering with the essential mathematical tools required for a comprehensive grasp of various theoretical physics disciplines. It introduces mathematical methods crucial for advanced theoretical physics courses requiring a firm background in tensor calculus. It begins with Manifolds in Chapter 1 by introducing the fundamental concept of manifolds, introducing curves, surfaces, and coordinate transformations. By introducing Einstein's summation convention, we explore mathematical expressions, paving the way for a concise exploration of length, area, and Volume in different geometries.

Chapter 2, which covers vector Calculus on Manifolds, takes us through the foundational understanding of scalars and vectors on a manifold. It unfolds the relationship between tangent vectors and basis vectors, providing the groundwork for a comprehensive exploration of vector calculus. The interplay between basis vectors, coordinate transformations, and the geometric insights derived from the metric tensor form the backbone of this chapter, culminating in an exploration of geodesics and the Euler-Lagrange equation.

Chapter 3 introduces tensor Calculus on Manifolds by exploring tensors and their multidimensional structures. It unravels the complexities of tensor components, symmetries, and transformations, shedding light on their geometric implications. Covariant and intrinsic derivatives take center stage, emphasizing their role in understanding the geometric properties of manifolds.

Finally, Chapter 4 shows the Tensor Application. In particular, we see the practical application of tensors in relativistic electrodynamics. This chapter navigates through the Lorentz force, electromagnetic field tensors, Maxwell's Equations, scalar and vector potentials, and the charged particle equation of motion within the framework of relativity.

Each chapter provides a set of problems that challenge readers to apply their understanding to real-world scenarios.

IOP Publishing

Studies in Theoretical Physics, Volume 2
Advanced mathematical methods
Daniel Erenso

Chapter 1

Manifolds

This chapter begins with the definition and explanation of a manifold, followed by an introduction to curves and surfaces, coordinate transformations in a manifold. After introducing Einstein's summation convention, a concise and elegant notation to express mathematical expressions more compactly, this chapter explores length, area, and volume in a manifold, discussing how these fundamental quantities are calculated on a manifold for different kinds of geometries. The chapter then discusses the concept of local Cartesian coordinates and tangent space, which is crucial in understanding vector calculus in a manifold, addressed in chapter two. The discussions of all these concepts are supported with several example problems and three-dimensional graphics. The chapter concludes with homework assignments to reinforce the concepts introduced.

1.1 What is a manifold?

To better understand the concept of a manifold, we begin by considering a ridged meterstick pinned at the north and south poles inside a hollow sphere as shown in figure 1.1. Initially at t_0 the sphere is at rest and begins to rotate. It is free to rotate about the x-axis, y-axis, and z-axis. We want to describe the angular position of the meterstick over a period of time, $t_0 \rightarrow t_1$, with a time interval of Δt. How many independent coordinates, which depend on time, do we need to describe the angular position of the meterstick relative to its initial position at t_0? Well, the answer is simple. We need *three* independent angular coordinates, known as *the Euler angles* $(\alpha(t),\ \beta(t),$ and $\gamma(t),$ introduced by Leonhard Euler to describe the orientation of a rigid body with respect to a fixed coordinate system), that are parameterized by time. Suppose we pick four instances, which we may describe by $t = t_0,\ t_0 + \Delta t,\ t_0 + 2\Delta t,$ and $t_1 = t_0 + 3\Delta t.$ Then over the time interval $t_0 \rightarrow t_1$ we have a set that consists of four points:

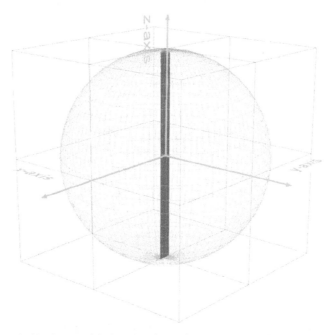

Figure 1.1. A ridged meterstick pinned at the north and south poles inside a hollow sphere.

$$\{[\alpha(t_0), \beta(t_0), \gamma(t_0)],$$
$$[\alpha(t_0 + \Delta t), \beta(t_0 + \Delta t), \gamma(t_0 + \Delta t)],$$
$$[\alpha(t_0 + 2\Delta t), \beta(t_0 + 2\Delta t), \gamma(t_0 + 2\Delta t)],$$
$$[\alpha(t_0 + 3\Delta t), \beta(t_0 + 3\Delta t), \gamma(t_0 + 3\Delta t)]\}, \qquad (1.1)$$

which describe the angular positions of the meterstick (see figure 1.2). We can make the time interval Δt infinitesimal to continuously describe the angular position of the meterstick. The resulting set of points forms a *manifold of dimension three*. A manifold can have different geometry (see figure 1.3).

Let us consider another example that we often see in classical mechanics, which is called *the phase space*. In phase space, we can describe the state of a particle over a period of time using the three coordinates of space describing the position of the particle and the three coordinates of velocity (or momentum) describing how fast the particle is moving at a given instant of time. In Cartesian coordinates, we use the coordinates $(x(t), y(t), z(t))$ for position and $(\dot{x}(t), \dot{y}(t), \dot{z}(t))$ for the velocity of the particle, which are parameterized by time, t. We can represent these six independent coordinates by $(x(t), y(t), z(t), \dot{x}(t), \dot{y}(t), \dot{z}(t))$. So when describing the state of a particle, say from $t = t_0$ to $t = t_1$, one can use an infinitesimal time interval so that one will have a set of points that can be parameterized continuously in terms of $(x(t), y(t), z(t), \dot{x}(t), \dot{y}(t), \dot{z}(t))$. These set of points form *a manifold of dimension six*.

Now let us apply this to the Minkowski space-time, where we have three space coordinates (x, y, z) and time t. This forms a four-dimensional manifold in general relativity with four coordinates each parametrized by the proper time τ

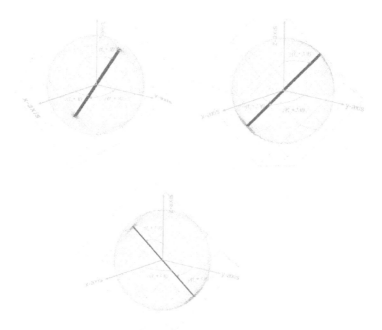

Figure 1.2. Angular position of the meterstick at $t_0 + \Delta t$, $t_0 + 2\Delta t$, $t_0 + 2\Delta t$, and $t_0 + 3\Delta t$.

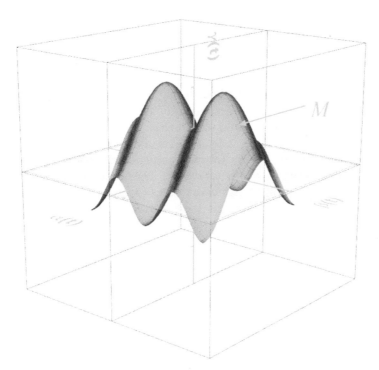

Figure 1.3. A curved surface in a three-dimensional manifold.

$(t(\tau),\ x(\tau),\ y(\tau),\ z(\tau))$. Therefore, an N-dimensional manifold M of points is one for which N independent real coordinates, which we represent by x^α $(x^1, x^2, x^3, \ldots, x^N)$, are required to specify any point completely. Note that it is understood that for $\alpha = 1, 2, 3\ldots N$ when we write x^α for an N-dimensional manifold. Thus for the three examples we considered above, we have for the angular position of the meterstick (three-dimensional manifold):

$$(\alpha(t),\ \beta(t),\ \gamma(t)) \rightarrow (x^1(t),\ x^2(t),\ x^3(t)), \tag{1.2}$$

and the phase-space (six-dimensional manifold):

$$(x(t),\ y(t),\ z(t),\ \dot{x}(t),\ \dot{y}(t),\ \dot{z}(t)) \rightarrow (x^1(t),\ x^2(t),\ x^3(t),\ x^4(t),\ x^5(t),\ x^6(t)), \tag{1.3}$$

and the Minkowski space-time (four-dimensional manifold):

$$(t(\tau),\ x(\tau),\ y(\tau),\ z(\tau)) \rightarrow (x^0(\tau),\ x^1(\tau),\ x^2(\tau),\ x^3(\tau)).$$

Note that for the Minkowski space-time manifold instead of $\alpha = 1, 2, 3, 4$ we use $\alpha = 0, 1, 2, 3$, where the first index is for the time coordinate and the last three indices are for the space coordinates.

Properties of a manifold

A manifold is continuous: Picking any point p in a manifold, one can find another point whose coordinates differ infinitesimally from point p.

A manifold is differentiable: Since a manifold is continuous, the derivative of a scalar function ϕ has a finite value at any point p in the manifold. That means for

$$\frac{d\phi}{dx^\alpha} = \lim_{\Delta x^\alpha \to 0} \left[\frac{\phi(x^\alpha + \Delta x^\alpha) - \phi(x^\alpha)}{\Delta x^\alpha} \right],$$

one finds a finite value.

Degeneracy in a manifold: Sometimes it may not be possible to cover the whole manifold with only one non-degenerate coordinate system. One example, is a plane which is a two-dimensional manifold (called R^2). A plane in polar coordinates has a degeneracy at the origin since φ is indeterminate at the origin as it can take any value in the domain, $0 \leqslant \varphi \leqslant 2\pi$. Figure 1.4 shows a two-dimensional plane manifold. In Cartesian coordinates (x, y), we do not see any degeneracy. However, if we attempt to map all these points using polar coordinates (ρ, φ), we will see that there is a degeneracy at the origin. Noting that the polar coordinates (ρ, φ) and the polar coordinates (x, y) are related by the equations

$$\rho^2 = x^2 + y^2,\ \varphi = \tan^{-1}\left(\frac{y}{x}\right). \tag{1.4}$$

Mapping of the plane in figure 1.4 for the coordinates

$$\rho = \pm\sqrt{x^2 + y^2},\ \varphi = \tan^{-1}\left(\frac{y}{x}\right), \tag{1.5}$$

where $-\pi \leqslant \varphi \leqslant \pi$, is shown in figure 1.5, wherein the region around $(0, \varphi)$ (where φ is indeterminate) is shown in white, which creates a discontinuity on this two-

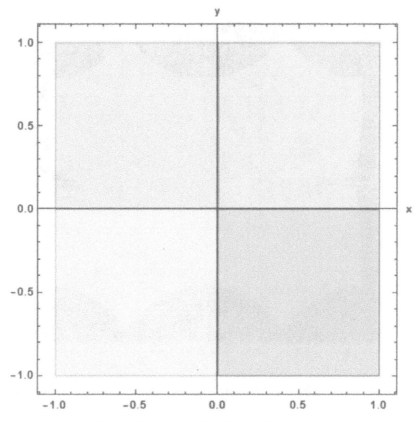

Figure 1.4. A plane in Cartesian coordinates (x, y).

dimensional manifold. Another example where we see degeneracy is on a spherical surface. A spherical surface is a two-dimensional manifold (called S^2). It can be described by two independent coordinates (φ, θ) except for two points on the manifold. These are the north and south pole where φ is indeterminate (or there is degeneracy) as shown in figure 1.6. Suppose the radius of the sphere is $r = 1$, the spherical (r, θ, φ) and the Cartesian (x, y, z) coordinates are related by

$$x = \sin(\theta)\cos(\varphi), \; y = \sin(\theta)\sin(\varphi) \Rightarrow \varphi = \tan^{-1}\left(\frac{y}{x}\right),$$

$$z = \cos(\theta) \Rightarrow x^2 + y^2 + z^2 = 1 \Rightarrow z = \pm\sqrt{1 - x^2 - y^2}$$

$$\Rightarrow \theta = \cos^{-1}\left[\pm\sqrt{1 - x^2 - y^2}\right] \tag{1.6}$$

$$\text{or } \theta = \tan^{-1}\left[\pm\frac{\sqrt{x^2 + y^2}}{\sqrt{1 - x^2 - y^2}}\right].$$

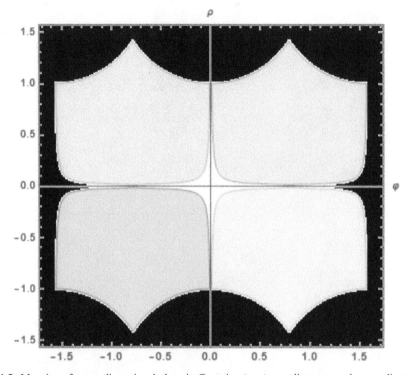

Figure 1.5. Mapping of a two-dimensional plane in Cartesian (x, y) coordinates to polar coordinates (ρ, φ) results in a discontinuous two-dimensional manifold.

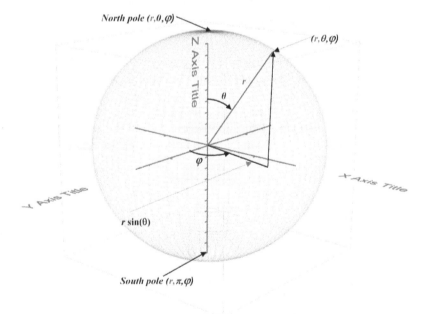

Figure 1.6. The north and south pole on the surface of sphere are degenerate in polar coordinates.

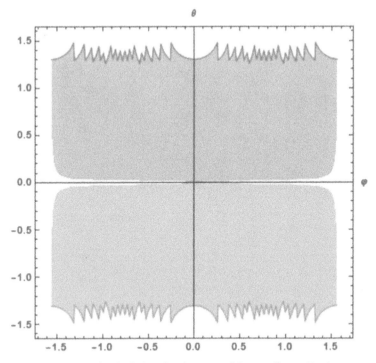

Figure 1.7. A spherical surface in terms of the coordinates (θ, φ).

Mapping of a spherical surface in terms of the coordinate (φ, θ) is shown in figure 1.7. It results in a discontinuous two-dimensional manifold due to the degeneracy at the north and south poles.

Coordinate patches
Coordinate patches are coordinate systems that cover a portion of the manifold where we have degeneracy. As shown in figure 1.5 and 1.7, no coordinate system covers the entire plane or sphere without running into the degenerate points. In this case, the smallest number of patches we need for the plane surface is one, whereas for the spherical surface is two.

Atlas
An atlas is a set of coordinate patches that covers the whole manifold.

1.2 Curves and surfaces in a manifold

Both curves and surfaces on a manifold are defined *parametrically*. That means we use some common parameters. For example, a curve in the phase space that we saw earlier, can be defined in terms of the time parameter t. Another example, a curve in the four-dimensional Minkowski space-time manifold is parameterized by the proper time and is defined by the interval

$$(ds(\tau))^2 = c^2 d\tau^2 = c^2 dt^2 - dx^2 - dy^2 - dz^2. \tag{1.7}$$

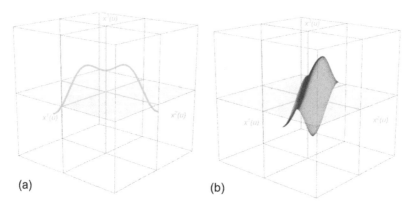

(a)   (b)

Figure 1.8. A curve (a) and a surface (b) in a three-dimensional manifold.

As we will see later, proper time is not a suitable parameter in the Minkowski space-time manifold all the time, in particular for photons. Thus generally, we use a parameter u to define a curve on a manifold.

A curve

A curve in a manifold of dimension N is defined by the parametric equation

$$x^\alpha = x^\alpha(u), \quad \text{where } \alpha = 1, 2, 3...N. \tag{1.8}$$

For example, the curve shown in figure 1.8(a) is defined by

$$x^1(u) = u, \quad x^2(u) = u^2, \quad x^3(u) = \sin^2(u) + \cos(u), \tag{1.9}$$

and it needs only one parameter u.

A surface

A surface in a manifold of dimension N (which is also referred as a submanifold) has M degrees of freedom (dimension) which is always less than the dimension of the manifold ($M < N$) and therefore it depends on M parameters which we represent as $u^1, u^2, u^3, ...u^M$ and is defined by the parametric equation

$$x^\alpha = x^\alpha(u^1, u^2, u^3, ...u^M), \quad \text{where } \alpha = 1, 2, 3...N. \tag{1.10}$$

A hypersurface

A surface of dimension M in a manifold of dimension N with $M = N - 1$. In this case the $N - 1$ parameters can be eliminated from the N equations and one can find the equation

$$f(x^1, x^2, x^3, ...x^N) = 0. \tag{1.11}$$

The surface shown in blue in figure 1.8(b) in the three-dimensional manifold needs two parameters to define it:

$$x^1 = u^1, \quad x^2 = u^2, \quad x^3 = \sin^2(u^1) + \cos(u^2). \tag{1.12}$$

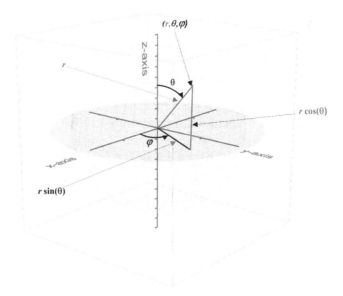

Figure 1.9. A spherical surface embedded in a three-dimensional manifold. The radius of the sphere is $r = a$. Adapted from [1]. © IOP Publishing Ltd. All rights reserved.

Note that this surface is in three-dimensional manifold and it is parameterized by two coordinates (u^1, u^2), ($M = N - 1 = 3 - 1 = 2$, it is a *hypersurface*).

Example 1.1. Let us consider the three-dimensional Euclidean manifold. A sphere is a highpersurface since $M = 2$, (figure 1.9). A point on the surface of a sphere is defined by

$$x^2 + y^2 + z^2 = a^2, \tag{1.13}$$

where a is the radius of the sphere. We note that in this case the surface of the sphere is a hypersurface that can be defined by the equation

$$g(x, y, z) = x^2 + y^2 + z^2 - a^2 = 0. \tag{1.14}$$

Note that (x, y, z) can be replaced by (x^1, x^2, x^3). Introducing the parameters (θ, φ, r) or (u_1, u_2, u_3) defined by

$$x = r \sin(\theta) \cos(\varphi), \; y = r \sin(\theta) \sin(\varphi), \; z = r \cos(\theta),$$

we can write the equations defining all points on the surface of a sphere with radius

$$ax = a \sin(\theta) \cos(\varphi), \; y = a \sin(\theta) \sin(\varphi), \; z = a \cos(\theta),$$

which shows it is parameterized by (θ, φ) or (u_1, u_2) (only $M = 2$ parameters) indicating that a spherical surface is a *hypersurface*. We can write the equation that define the surface of the sphere ($M = 2$) that is embedded in a three-dimensional manifold ($N = 3$) using only one parameter (by eliminating the $N - 1 = 2$, parameters) as

$$g(\theta, \varphi, r) = [r \sin(\theta) \cos(\varphi)]^2 + [r \sin(\theta) \sin(\varphi)]^2$$
$$+ [r \cos(\theta)]^2 - a^2 = 0 \qquad (1.15)$$
$$\Rightarrow g(\theta, \varphi, r) = r^2 - a^2 = 0,$$

which is the property of a hypersurface in a manifold.

Therefore if there is a point on a hypersurface ($M = N - 1$ dimensional submanifold embedded in N-dimensional manifold), then the point coordinate must satisfy equation (1.11). We come up with a similar generalization to this for a point that belong to any surface with dimension M in a manifold of dimension N ($M < N$) must satisfy the constraints

$$g_1(x^1, x^2, x^3, \ldots x^N) = 0, \; g_2(x^1, x^2, x^3, \ldots x^N) = 0,$$
$$g_3(x^1, x^2, x^3, \ldots x^N) = 0, \; \ldots g_{N-M}(x^1, x^2, x^3, \ldots x^N) = 0. \qquad (1.16)$$

àFor example, for a surface with dimension $M = 2$ embedded in a manifold with dimension $N = 4$ (which is clearly not a hypersurface), we find

$$g_1(x^1, x^2, x^3, x^4) = 0, \; g_2(x^1, x^2, x^3, x^4) = 0. \qquad (1.17)$$

This means if a point belongs to the surface with dimension $M = 2$ in a manifold $N = 4$, it must satisfy these two constraints.

1.3 Coordinate transformations and summation convention

Let us consider the three-dimensional Euclidean manifold. A point in this manifold can be represented using Cartesian coordinates (x, y, z) which we shall represent by (x'^1, x'^2, x'^3). This same point can also be represented using spherical coordinates (r, θ, φ) (see figure 1.10) that we represent by (x^1, x^2, x^3). Now the question is how we relate the Cartesian coordinates with the spherical coordinates or vice versa. In terms of these notations, one can write

$$r \to r(x, y, z), \; \text{or } x^1 \to x^1(x'^1, x'^2, x'^3),$$
$$\theta \to \theta(x, y, z), \; \text{or } x^2 \to x^2(x'^1, x'^2, x'^3), \qquad (1.18)$$
$$\varphi \to \varphi(x, y, z), \; \text{or } x^3 \to x^3(x'^1, x'^2, x'^3),$$

or

$$x \to x(r, \theta, \varphi), \; \text{or } x'^1 \to x'^1(x^1, x^2, x^3),$$
$$y \to y(r, \theta, \varphi), \; \text{or } x'^2 \to x'^2(x^1, x^2, x^3), \qquad (1.19)$$
$$z \to z(r, \theta, \varphi), \; \text{or } x'^3 \to x'^3(x^1, x^2, x^3).$$

Now let us consider the function $g(x, y, z)$ which we represent by $g(r, \theta, \varphi)$. These function can be represented, in terms of the notations introduced above, by

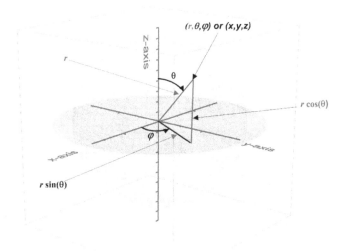

Figure 1.10. Cartesian (x, y, z) and spherical coordinates (r, θ, φ). Adapted from [xx]. © IOP Publishing Ltd.

$g(x^{'1}, x^{'2}, x^{'3})$ and $g(x^1, x^2, x^3)$, respectively. As an example, consider the electric potential at a position $\vec{r} = (x, y, z) = (r, \theta, \varphi)$ due to a point charge, q, located at the position $\vec{r}_0 = (x_0, y_0, z_0) = (r_0, \theta_0, \varphi_0)$ as shown in figure 1.11. This electric potential (which we represent by $g(x, y, z)$), in terms of the Cartesian coordinates (x, y, z) is given by

$$g(x, y, z) = \frac{1}{4\pi\epsilon_0} \frac{q}{|\vec{r} - \vec{r}_0|} = -\frac{1}{4\pi\epsilon_0} \frac{q}{\sqrt{(x - x_0)^2 + (x - y_0)^2 + (x - z_0)^2}}, \quad (1.20)$$

or in terms of the spherical coordinates (r, θ, φ) by

$$g(r, \theta, \varphi) = \frac{1}{4\pi\epsilon_0} \frac{q}{\sqrt{r^2 + r_0^2 + 2rr_0 \cos(\gamma(\theta, \varphi))}}, \quad (1.21)$$

where

$$\gamma(\theta, \varphi) = \cos(\theta)\cos(\theta') + \sin(\theta)\sin(\theta')\cos(\varphi - \varphi'). \quad (1.22)$$

Now let us say we are interested in the electric field $\vec{E}$ due to this point charge which is given by

$$\vec{E} = -\nabla g. \quad (1.23)$$

For the sake of simplicity if we drop the negative sign, one can write the electric field vector in the Cartesian coordinates (x, y, z) as

$$\vec{E}(x, y, z) = \frac{\partial g}{\partial x}\hat{x} + \frac{\partial g}{\partial y}\hat{y} + \frac{\partial g}{\partial z}\hat{z}.$$

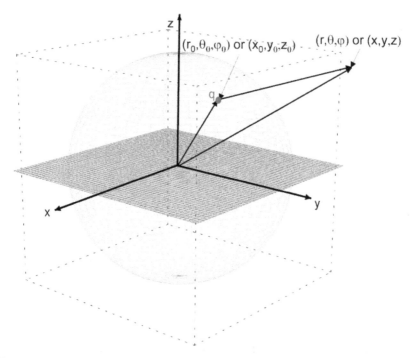

Figure 1.11. A point charge q located at $(r_0, \theta_0, \varphi_0)$ and the electric potential at point (r, θ, φ).

If the explicit form of the electrical potential is known in terms of the spherical coordinates, (r, θ, φ), then to find the electric field in Cartesian coordinates, we can write

$$
\frac{\partial g}{\partial x} = \frac{\partial g}{\partial r}\frac{\partial r}{\partial x} + \frac{\partial g}{\partial \theta}\frac{\partial \theta}{\partial x} + \frac{\partial g}{\partial \varphi}\frac{\partial \varphi}{\partial x}
$$

$$
\frac{\partial g}{\partial y} = \frac{\partial g}{\partial r}\frac{\partial r}{\partial y} + \frac{\partial g}{\partial \theta}\frac{\partial \theta}{\partial y} + \frac{\partial g}{\partial \varphi}\frac{\partial \varphi}{\partial y}, \qquad (1.24)
$$

$$
\frac{\partial g}{\partial z} = \frac{\partial g}{\partial r}\frac{\partial r}{\partial z} + \frac{\partial g}{\partial \theta}\frac{\partial \theta}{\partial z} + \frac{\partial g}{\partial \varphi}\frac{\partial \varphi}{\partial z},
$$

Using matrices, equation (1.24), can be put in the form

$$
\begin{bmatrix} \dfrac{\partial g}{\partial x} \\[2ex] \dfrac{\partial g}{\partial y} \\[2ex] \dfrac{\partial g}{\partial z} \end{bmatrix} = \begin{bmatrix} \dfrac{\partial r}{\partial x} & \dfrac{\partial \theta}{\partial x} & \dfrac{\partial \varphi}{\partial x} \\[2ex] \dfrac{\partial r}{\partial y} & \dfrac{\partial \theta}{\partial y} & \dfrac{\partial \varphi}{\partial y} \\[2ex] \dfrac{\partial r}{\partial z} & \dfrac{\partial \theta}{\partial z} & \dfrac{\partial \varphi}{\partial z} \end{bmatrix} \begin{bmatrix} \dfrac{\partial g}{\partial r} \\[2ex] \dfrac{\partial g}{\partial \theta} \\[2ex] \dfrac{\partial g}{\partial \varphi} \end{bmatrix}, \qquad (1.25)
$$

or in terms of the notation introduced,

$$
\begin{bmatrix} \dfrac{\partial g}{\partial x'^1} \\[2ex] \dfrac{\partial g}{\partial x'^2} \\[2ex] \dfrac{\partial g}{\partial x'^3} \end{bmatrix} = \begin{bmatrix} \dfrac{\partial x^1}{\partial x'^1} & \dfrac{\partial x^2}{\partial x'^1} & \dfrac{\partial x^3}{\partial x'^1} \\[2ex] \dfrac{\partial x^1}{\partial x'^2} & \dfrac{\partial x^2}{\partial x'^2} & \dfrac{\partial x^3}{\partial x'^2} \\[2ex] \dfrac{\partial x^1}{\partial x'^3} & \dfrac{\partial x^2}{\partial x'^3} & \dfrac{\partial x^3}{\partial x'^3} \end{bmatrix} \begin{bmatrix} \dfrac{\partial g}{\partial x^1} \\[2ex] \dfrac{\partial g}{\partial x^2} \\[2ex] \dfrac{\partial g}{\partial x^3} \end{bmatrix}. \tag{1.26}
$$

Similarly, for the partial derivative of the function $g(x, y, z)$ with respect to r, θ, and φ, we can write

$$
\begin{aligned}
\frac{\partial g}{\partial r} &= \frac{\partial g}{\partial x}\frac{\partial x}{\partial r} + \frac{\partial g}{\partial y}\frac{\partial y}{\partial r} + \frac{\partial g}{\partial z}\frac{\partial z}{\partial r} \\[1ex]
\frac{\partial g}{\partial \theta} &= \frac{\partial g}{\partial x}\frac{\partial x}{\partial \theta} + \frac{\partial g}{\partial y}\frac{\partial y}{\partial \theta} + \frac{\partial g}{\partial z}\frac{\partial z}{\partial \theta}, \\[1ex]
\frac{\partial g}{\partial \varphi} &= \frac{\partial g}{\partial r}\frac{\partial r}{\partial \varphi} + \frac{\partial g}{\partial \theta}\frac{\partial \theta}{\partial \varphi} + \frac{\partial g}{\partial \varphi}\frac{\partial \varphi}{\partial \varphi},
\end{aligned} \tag{1.27}
$$

which can also be put, using matrices, as

$$
\begin{bmatrix} \dfrac{\partial g}{\partial r} \\[2ex] \dfrac{\partial g}{\partial \theta} \\[2ex] \dfrac{\partial g}{\partial \varphi} \end{bmatrix} = \begin{bmatrix} \dfrac{\partial x}{\partial r} & \dfrac{\partial y}{\partial r} & \dfrac{\partial z}{\partial r} \\[2ex] \dfrac{\partial x}{\partial \theta} & \dfrac{\partial y}{\partial \theta} & \dfrac{\partial z}{\partial \theta} \\[2ex] \dfrac{\partial x}{\partial \varphi} & \dfrac{\partial y}{\partial \varphi} & \dfrac{\partial z}{\partial \varphi} \end{bmatrix} \begin{bmatrix} \dfrac{\partial g}{\partial x} \\[2ex] \dfrac{\partial g}{\partial y} \\[2ex] \dfrac{\partial g}{\partial z} \end{bmatrix},
$$

$$
\Rightarrow \begin{bmatrix} \dfrac{\partial g}{\partial x^1} \\[2ex] \dfrac{\partial g}{\partial x^2} \\[2ex] \dfrac{\partial g}{\partial x^3} \end{bmatrix} = \begin{bmatrix} \dfrac{\partial x'^1}{\partial x^1} & \dfrac{\partial x'^2}{\partial x^1} & \dfrac{\partial x'^3}{\partial x^1} \\[2ex] \dfrac{\partial x'^1}{\partial x^2} & \dfrac{\partial x'^2}{\partial x^2} & \dfrac{\partial x'^3}{\partial x^2} \\[2ex] \dfrac{\partial x'^1}{\partial x^3} & \dfrac{\partial x'^2}{\partial x^3} & \dfrac{\partial x'^3}{\partial x^3} \end{bmatrix} \begin{bmatrix} \dfrac{\partial g}{\partial x'^1} \\[2ex] \dfrac{\partial g}{\partial x'^2} \\[2ex] \dfrac{\partial g}{\partial x'^3} \end{bmatrix}. \tag{1.28}
$$

Using equation (1.28), we can rewrite equation (1.26) as

$$
\begin{bmatrix} \dfrac{\partial g}{\partial x'^1} \\[2ex] \dfrac{\partial g}{\partial x'^2} \\[2ex] \dfrac{\partial g}{\partial x'^3} \end{bmatrix} = \begin{bmatrix} \dfrac{\partial x^1}{\partial x'^1} & \dfrac{\partial x^2}{\partial x'^1} & \dfrac{\partial x^3}{\partial x'^1} \\[2ex] \dfrac{\partial x^1}{\partial x'^2} & \dfrac{\partial x^2}{\partial x'^2} & \dfrac{\partial x^3}{\partial x'^2} \\[2ex] \dfrac{\partial x^1}{\partial x'^3} & \dfrac{\partial x^2}{\partial x'^3} & \dfrac{\partial x^3}{\partial x'^3} \end{bmatrix} \begin{bmatrix} \dfrac{\partial x'^1}{\partial x^1} & \dfrac{\partial x'^2}{\partial x^1} & \dfrac{\partial x'^3}{\partial x^1} \\[2ex] \dfrac{\partial x'^1}{\partial x^2} & \dfrac{\partial x'^2}{\partial x^2} & \dfrac{\partial x'^3}{\partial x^2} \\[2ex] \dfrac{\partial x'^1}{\partial x^3} & \dfrac{\partial x'^2}{\partial x^3} & \dfrac{\partial x'^3}{\partial x^3} \end{bmatrix} \begin{bmatrix} \dfrac{\partial g}{\partial x'^1} \\[2ex] \dfrac{\partial g}{\partial x'^2} \\[2ex] \dfrac{\partial g}{\partial x'^3} \end{bmatrix}. \tag{1.29}
$$

The left and the right sides of equation (1.29) are equal if

$$\begin{bmatrix} \dfrac{\partial x^1}{\partial x'^1} & \dfrac{\partial x^2}{\partial x'^1} & \dfrac{\partial x^3}{\partial x'^1} \\[2mm] \dfrac{\partial x^1}{\partial x'^2} & \dfrac{\partial x^2}{\partial x'^2} & \dfrac{\partial x^3}{\partial x'^2} \\[2mm] \dfrac{\partial x^1}{\partial x'^3} & \dfrac{\partial x^2}{\partial x'^3} & \dfrac{\partial x^3}{\partial x'^3} \end{bmatrix} \begin{bmatrix} \dfrac{\partial x'^1}{\partial x^1} & \dfrac{\partial x'^2}{\partial x^1} & \dfrac{\partial x'^3}{\partial x^1} \\[2mm] \dfrac{\partial x'^1}{\partial x^2} & \dfrac{\partial x'^2}{\partial x^2} & \dfrac{\partial x'^3}{\partial x^2} \\[2mm] \dfrac{\partial x'^1}{\partial x^3} & \dfrac{\partial x'^2}{\partial x^3} & \dfrac{\partial x'^3}{\partial x^3} \end{bmatrix} = \begin{bmatrix} 1 & 0 & 0 \\ 0 & 1 & 0 \\ 0 & 0 & 1 \end{bmatrix}. \tag{1.30}$$

This clearly indicates that the transformation can exist provided for the matrix

$$A = \begin{bmatrix} \dfrac{\partial x'^1}{\partial x^1} & \dfrac{\partial x'^2}{\partial x^1} & \dfrac{\partial x'^3}{\partial x^1} \\[2mm] \dfrac{\partial x'^1}{\partial x^2} & \dfrac{\partial x'^2}{\partial x^2} & \dfrac{\partial x'^3}{\partial x^2} \\[2mm] \dfrac{\partial x'^1}{\partial x^3} & \dfrac{\partial x'^2}{\partial x^3} & \dfrac{\partial x'^3}{\partial x^3} \end{bmatrix}. \tag{1.31}$$

The matrix

$$A^{-1} = \begin{bmatrix} \dfrac{\partial x^1}{\partial x'^1} & \dfrac{\partial x^2}{\partial x'^1} & \dfrac{\partial x^3}{\partial x'^1} \\[2mm] \dfrac{\partial x^1}{\partial x'^2} & \dfrac{\partial x^2}{\partial x'^2} & \dfrac{\partial x^3}{\partial x'^2} \\[2mm] \dfrac{\partial x^1}{\partial x'^3} & \dfrac{\partial x^2}{\partial x'^3} & \dfrac{\partial x^3}{\partial x'^3} \end{bmatrix}, \tag{1.32}$$

is an inverse matrix so that

$$A^{-1}A = AA^{-1} = I, \tag{1.33}$$

where I is the identity matrix:

$$I = \begin{bmatrix} 1 & 0 & 0 \\ 0 & 1 & 0 \\ 0 & 0 & 1 \end{bmatrix}.$$

This means the transformation from the non-prime (spherical) to the prime (Cartesian) coordinates,

$$\begin{bmatrix} \dfrac{\partial g}{\partial x'^1} \\[2mm] \dfrac{\partial g}{\partial x'^2} \\[2mm] \dfrac{\partial g}{\partial x'^3} \end{bmatrix} = \begin{bmatrix} \dfrac{\partial x^1}{\partial x'^1} & \dfrac{\partial x^2}{\partial x'^1} & \dfrac{\partial x^3}{\partial x'^1} \\[2mm] \dfrac{\partial x^1}{\partial x'^2} & \dfrac{\partial x^2}{\partial x'^2} & \dfrac{\partial x^3}{\partial x'^2} \\[2mm] \dfrac{\partial x^1}{\partial x'^3} & \dfrac{\partial x^2}{\partial x'^3} & \dfrac{\partial x^3}{\partial x'^3} \end{bmatrix} \begin{bmatrix} \dfrac{\partial g}{\partial x^1} \\[2mm] \dfrac{\partial g}{\partial x^2} \\[2mm] \dfrac{\partial g}{\partial x^3} \end{bmatrix}, \tag{1.34}$$

can be rewritten as

$$
\begin{bmatrix} \dfrac{\partial g}{\partial x'^1} \\[2mm] \dfrac{\partial g}{\partial x'^2} \\[2mm] \dfrac{\partial g}{\partial x'^3} \end{bmatrix} = A^{-1} \begin{bmatrix} \dfrac{\partial g}{\partial x^1} \\[2mm] \dfrac{\partial g}{\partial x^2} \\[2mm] \dfrac{\partial g}{\partial x^3} \end{bmatrix}.
\tag{1.35}
$$

Orthogonal transformation: in an orthogonal transformation of coordinates the length (magnitude) of the vector (the electric field vector in this particular case) is invariant. This means independent of the coordinates system, we must find the same value for the magnitude of the vector. This leads to one important condition for such coordinate transformation. Next, we derive this condition using a two-dimensional vector with components (R_x, R_y) and (R_x', R_y') in two different coordinate systems. Suppose these components are related by the matrix equation

$$
\begin{bmatrix} R_x \\ R_y \end{bmatrix} = A \begin{bmatrix} R_x' \\ R_y' \end{bmatrix} \Rightarrow \begin{bmatrix} R_x' \\ R_y' \end{bmatrix} = A^{-1} \begin{bmatrix} R_x \\ R_y \end{bmatrix},
\tag{1.36}
$$

upon taking the transpose of both sides, we have

$$
\begin{bmatrix} R_x \\ R_y \end{bmatrix}^T = \begin{bmatrix} R_x' \\ R_y' \end{bmatrix}^T A^T \Rightarrow \begin{bmatrix} R_x & R_y \end{bmatrix} = \begin{bmatrix} R_x' & R_y' \end{bmatrix} A^T.
\tag{1.37}
$$

so that

$$
\begin{bmatrix} R_x & R_y \end{bmatrix} \begin{bmatrix} R_x \\ R_y \end{bmatrix} = \begin{bmatrix} R_x & R_y \end{bmatrix} A^T A \begin{bmatrix} R_x' \\ R_y' \end{bmatrix}.
\tag{1.38}
$$

For the magnitude of the vector to be invariant,

$$
\begin{bmatrix} R_x & R_y \end{bmatrix} \begin{bmatrix} R_x \\ R_y \end{bmatrix} = \begin{bmatrix} R_x & R_y \end{bmatrix} \begin{bmatrix} R_x' \\ R_y' \end{bmatrix},
\tag{1.39}
$$

we must then have

$$
A^T A = I \Rightarrow A^T = A^{-1}
\tag{1.40}
$$

for an orthogonal coordinate transformation.

In view of this condition for an orthogonal transformation, one can rewrite equation (1.35) as

$$
\begin{bmatrix} \dfrac{\partial g}{\partial x'^1} \\[2mm] \dfrac{\partial g}{\partial x'^2} \\[2mm] \dfrac{\partial g}{\partial x'^3} \end{bmatrix} = A^T \begin{bmatrix} \dfrac{\partial g}{\partial x^1} \\[2mm] \dfrac{\partial g}{\partial x^2} \\[2mm] \dfrac{\partial g}{\partial x^3} \end{bmatrix},
\tag{1.41}
$$

where A^{T} is given by

$$
A^{\mathrm{T}} = \begin{bmatrix} \dfrac{\partial x'^1}{\partial x^1} & \dfrac{\partial x'^2}{\partial x^1} & \dfrac{\partial x'^3}{\partial x^1} \\[2mm] \dfrac{\partial x'^1}{\partial x^2} & \dfrac{\partial x'^2}{\partial x^2} & \dfrac{\partial x'^3}{\partial x^2} \\[2mm] \dfrac{\partial x'^1}{\partial x^3} & \dfrac{\partial x'^2}{\partial x^3} & \dfrac{\partial x'^3}{\partial x^3} \end{bmatrix}^{\mathrm{T}} = \begin{bmatrix} \dfrac{\partial x'^1}{\partial x^1} & \dfrac{\partial x'^1}{\partial x^2} & \dfrac{\partial x'^1}{\partial x^3} \\[2mm] \dfrac{\partial x'^2}{\partial x^1} & \dfrac{\partial x'^2}{\partial x^2} & \dfrac{\partial x'^2}{\partial x^2} \\[2mm] \dfrac{\partial x'^3}{\partial x^1} & \dfrac{\partial x'^3}{\partial x^2} & \dfrac{\partial x'^3}{\partial x^3} \end{bmatrix} = \begin{bmatrix} \dfrac{\partial x'^\alpha}{\partial x^\beta} \end{bmatrix}. \tag{1.42}
$$

Similarly, for the inverse transformation, one can write

$$
A \begin{bmatrix} \dfrac{\partial g}{\partial x'^1} \\[2mm] \dfrac{\partial g}{\partial x'^2} \\[2mm] \dfrac{\partial g}{\partial x'^3} \end{bmatrix} = A A^{-1} \begin{bmatrix} \dfrac{\partial g}{\partial x^1} \\[2mm] \dfrac{\partial g}{\partial x^2} \\[2mm] \dfrac{\partial g}{\partial x^3} \end{bmatrix} \Rightarrow \begin{bmatrix} \dfrac{\partial g}{\partial x^1} \\[2mm] \dfrac{\partial g}{\partial x^2} \\[2mm] \dfrac{\partial g}{\partial x^3} \end{bmatrix} = A \begin{bmatrix} \dfrac{\partial g}{\partial x'^1} \\[2mm] \dfrac{\partial g}{\partial x'^2} \\[2mm] \dfrac{\partial g}{\partial x'^3} \end{bmatrix}, \tag{1.43}
$$

where the transformation matrix is given by

$$
A = \begin{bmatrix} \dfrac{\partial x'^\alpha}{\partial x^\beta} \end{bmatrix}^{\mathrm{T}}. \tag{1.44}
$$

For a manifold of dimension N, the transformation matrix is given by

$$
\frac{\partial x'^\alpha}{\partial x^\beta} = \begin{bmatrix} \dfrac{\partial x'^1}{\partial x^1} & \dfrac{\partial x'^1}{\partial x^2} & \cdots & \dfrac{\partial x'^1}{\partial x^N} \\[2mm] \dfrac{\partial x'^2}{\partial x^1} & \dfrac{\partial x'^2}{\partial x^2} & \cdots & \dfrac{\partial x'^2}{\partial x^3} \\[2mm] \vdots & \vdots & \vdots & \vdots \\[2mm] \dfrac{\partial x'^N}{\partial x^1} & \dfrac{\partial x'^N}{\partial x^2} & \cdots & \dfrac{\partial x'^N}{\partial x^N} \end{bmatrix}. \tag{1.45}
$$

As we recall, a matrix is invertible provided its determinant is different from zero. Therefore, the inverse transformation is possible when the determinant of the transformation matrix, which is known as *the Jacobian*, J, is different from zero:

$$
J = \det \begin{bmatrix} \dfrac{\partial x'^\alpha}{\partial x^\beta} \end{bmatrix} = \begin{vmatrix} \dfrac{\partial x'^1}{\partial x^1} & \dfrac{\partial x'^1}{\partial x^2} & \cdots & \dfrac{\partial x'^1}{\partial x^N} \\[2mm] \dfrac{\partial x'^2}{\partial x^1} & \dfrac{\partial x'^2}{\partial x^2} & \cdots & \dfrac{\partial x'^2}{\partial x^3} \\[2mm] \vdots & \vdots & \vdots \\[2mm] \dfrac{\partial x'^N}{\partial x^1} & \dfrac{\partial x'^N}{\partial x^2} & \cdots & \dfrac{\partial x'^N}{\partial x^N} \end{vmatrix} \neq 0. \tag{1.46}
$$

The inverse transformation matrix can be written as

$$
\left[\frac{\partial x^\alpha}{\partial x'^\beta} \right] =
\begin{bmatrix}
\dfrac{\partial x^1}{\partial x'^1} & \dfrac{\partial x^1}{\partial x'^2} & \cdots & \dfrac{\partial x^1}{\partial x'^N} \\[2mm]
\dfrac{\partial x^2}{\partial x'^1} & \dfrac{\partial x^2}{\partial x'^2} & \vdots & \dfrac{\partial x^2}{\partial x'^3} \\[2mm]
\vdots & \vdots & \vdots & \vdots \\[2mm]
\dfrac{\partial x^N}{\partial x'^1} & \dfrac{\partial x^N}{\partial x'^2} & \cdots & \dfrac{\partial x^N}{\partial x'^N}
\end{bmatrix},
\tag{1.47}
$$

and the Jacobian J'

$$
J' = \det \left[\frac{\partial x^\alpha}{\partial x'^\beta} \right] =
\begin{vmatrix}
\dfrac{\partial x^1}{\partial x'^1} & \dfrac{\partial x^1}{\partial x'^2} & \cdots & \dfrac{\partial x^1}{\partial x'^N} \\[2mm]
\dfrac{\partial x^2}{\partial x'^1} & \dfrac{\partial x^2}{\partial x'^2} & \vdots & \dfrac{\partial x^2}{\partial x'^3} \\[2mm]
\vdots & \vdots & \vdots & \vdots \\[2mm]
\dfrac{\partial x^N}{\partial x'^1} & \dfrac{\partial x^N}{\partial x'^2} & \cdots & \dfrac{\partial x^N}{\partial x'^N}
\end{vmatrix}.
\tag{1.48}
$$

Example 1.2. A rotation of the Cartesian coordinates (x, y, z) about the z-axis gives another coordinate system, (X, Y, Z):

$$
\begin{aligned}
X &= x \cos(\theta) + y \sin(\theta), \\
Y &= -x \sin(\theta) + y \cos(\theta), \\
Z &= z.
\end{aligned}
\tag{1.49}
$$

(See figure 1.12).

(a) Using the given notation $(x, y, z) \to (x'^1, x'^2, x'^3)$ and $(X, Y, Z) \to (x^1, x^2, x^3)$, find the transformation matrix

$$
\left[\frac{\partial x^\alpha}{\partial x'^\beta} \right],
\tag{1.50}
$$

and the Jacobian, J,

$$
J = \det \left[\frac{\partial x^\alpha}{\partial x'^\beta} \right].
\tag{1.51}
$$

(b) Find the inverse matrix

$$
\left[\frac{\partial x^\alpha}{\partial x'^\beta} \right]^{-1}.
\tag{1.52}
$$

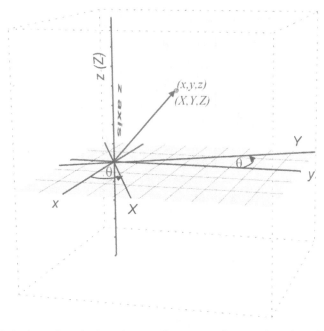

Figure 1.12. A rotation of a Cartesian coordinate system by an angle θ about the z-axis.

(c) Show that

$$\left[\frac{\partial x^{'\alpha}}{\partial x^{\beta}}\right] = \left[\frac{\partial x^{\alpha}}{\partial x^{'\beta}}\right]^{\mathrm{T}}.$$ (1.53)

Solution:

(a) For the given notation, we can rewrite equation (1.49) as

$$x^1 = x^{'1}\cos(\theta) + x^{'2}\sin(\theta),$$
$$x^2 = -x^{'1}\sin(\theta) + x^{'2}\cos(\theta),$$ (1.54)
$$x^3 = x^{'3},$$

so that for

$$\frac{\partial x^{\alpha}}{\partial x^{'\beta}}$$ (1.55)

one finds

$$\frac{\partial x^1}{\partial x^{'1}} = \cos(\theta), \quad \frac{\partial x^1}{\partial x^{'2}} = \sin(\theta), \quad \frac{\partial x^1}{\partial x^{'3}} = 0$$

$$\frac{\partial x^2}{\partial x^{'1}} = -\sin(\theta), \quad \frac{\partial x^2}{\partial x^{'2}} = \cos(\theta), \quad \frac{\partial x^2}{\partial x^{'3}} = 0$$

$$\frac{\partial x^3}{\partial x^{'1}} = 0, \quad \frac{\partial x^3}{\partial x^{'2}} = 0, \quad \frac{\partial x^3}{\partial x^{'3}} = 1,$$

and the transformation matrix becomes

$$\frac{\partial x^{\alpha}}{\partial x'^{\beta}} = \begin{vmatrix} \dfrac{\partial x'^1}{\partial x^1} & \dfrac{\partial x'^1}{\partial x^2} & \dfrac{\partial x'^1}{\partial x^3} \\[2mm] \dfrac{\partial x'^2}{\partial x^1} & \dfrac{\partial x'^2}{\partial x^2} & \dfrac{\partial x'^2}{\partial x^2} \\[2mm] \dfrac{\partial x'^3}{\partial x^1} & \dfrac{\partial x'^3}{\partial x^2} & \dfrac{\partial x'^3}{\partial x^3} \end{vmatrix} = \begin{bmatrix} \cos(\theta) & \sin(\theta) & 0 \\ -\sin(\theta) & \cos(\theta) & 0 \\ 0 & 0 & 1 \end{bmatrix},$$

which is the rotation matrix $R(\theta) = R$. The Jacobian of this transformation matrix is given by

$$J = \det\left[\frac{\partial x'^a}{\partial x^{\beta}}\right] = \det \begin{vmatrix} \cos(\theta) & \sin(\theta) & 0 \\ -\sin(\theta) & \cos(\theta) & 0 \\ 0 & 0 & 1 \end{vmatrix} = 1,$$

which is different from zero.

(b) We recall (from volume one, section 4.7), the inverse for a matrix

$$A = \left[\frac{\partial x'^a}{\partial x^{\beta}}\right] = \begin{bmatrix} \cos(\theta) & \sin(\theta) & 0 \\ -\sin(\theta) & \cos(\theta) & 0 \\ 0 & 0 & 1 \end{bmatrix}$$

is given by

$$A^{-1} = \frac{[cof(A)]^{\mathrm{T}}}{\det |A|}, \tag{1.56}$$

so that

$$[cof(A)]^{\mathrm{T}} = \begin{bmatrix} \cos(\theta) & -\sin(\theta) & 0 \\ \sin(\theta) & \cos(\theta) & 0 \\ 0 & 0 & 1 \end{bmatrix}.$$

Noting that the det $|A| = 1$, one finds for the inverse matrix

$$\left[\frac{\partial x^{\alpha}}{\partial x'^{\beta}}\right]^{-1} = \begin{bmatrix} \cos(\theta) & -\sin(\theta) & 0 \\ \sin(\theta) & \cos(\theta) & 0 \\ 0 & 0 & 1 \end{bmatrix} = \left[\frac{\partial x'^a}{\partial x^{\beta}}\right]. \tag{1.57}$$

(c) Noting that the transpose of the transformation matrix

$$\left[\frac{\partial x^{\alpha}}{\partial x'^{\beta}}\right]^{\mathrm{T}} = \begin{bmatrix} \cos(\theta) & -\sin(\theta) & 0 \\ \sin(\theta) & \cos(\theta) & 0 \\ 0 & 0 & 1 \end{bmatrix},$$

one can easily see that

$$\left[\frac{\partial x^{'\alpha}}{\partial x^{\beta}} \right] = \left[\frac{\partial x^{\alpha}}{\partial x^{'\beta}} \right]^{\mathrm{T}}. \tag{1.58}$$

We note that

$$\frac{\partial x^{'\alpha}}{\partial x^{'1}} = \frac{\partial x^{'\alpha}}{\partial x^{1}} \frac{\partial x^{1}}{\partial x^{'1}} + \frac{\partial x^{'\alpha}}{\partial x^{2}} \frac{\partial x^{2}}{\partial x^{'1}} + \cdots \frac{\partial x^{'\alpha}}{\partial x^{N}} \frac{\partial x^{N}}{\partial x^{'1}} = \sum_{\beta=1}^{N} \frac{\partial x^{'\alpha}}{\partial x^{\beta}} \frac{\partial x^{\beta}}{\partial x^{'1}}. \tag{1.59}$$

Noting that for independent coordinates

$$\frac{\partial x^{'\alpha}}{\partial x^{'\kappa}} = \begin{cases} 0, & \alpha \neq \kappa, \\ 1, & \alpha = \kappa, \end{cases} \tag{1.60}$$

we can generally write

$$\sum_{\beta=1}^{N} \frac{\partial x^{'\alpha}}{\partial x^{\beta}} \frac{\partial x^{\beta}}{\partial x^{'\kappa}} = \delta_{\kappa}^{\alpha}. \tag{1.61}$$

Consider two points, P and Q, in a manifold with dimension N. Suppose these points are separated by an infinitesimal interval so that if the coordinates of P is x^{α} and that of Q is $x^{\alpha} + dx^{\alpha}$, then one can write

$$dx^{'\alpha} = \frac{\partial x^{'\alpha}}{\partial x^{1}} dx^{1} + \frac{\partial x^{'\alpha}}{\partial x^{2}} dx^{2} \cdots \frac{\partial x^{'\alpha}}{\partial x^{N}} dx^{N} = \sum_{\beta=1}^{N} \frac{\partial x^{'\alpha}}{\partial x^{\beta}} dx^{\beta}, \tag{1.62}$$

where the summation is evaluated at P. Similarly, for the interval between $x^{'\alpha}$ and $x^{'\alpha} + dx^{'\alpha}$, in the unprimed coordinate system, we can write

$$dx^{\alpha} = \frac{\partial x^{\alpha}}{\partial x^{'1}} dx^{'1} + \frac{\partial x^{\alpha}}{\partial x^{'2}} dx^{'2} \cdots \frac{\partial x^{\alpha}}{\partial x^{'N}} dx^{'N} = \sum_{\beta=1}^{N} \frac{\partial x^{\alpha}}{\partial x^{'\beta}} dx^{'\beta}, \tag{1.63}$$

where also the summation is evaluated at P (figure 1.13).

Einstein's summation convention: whenever an index occurs twice in an expression, once as subscript and once as a superscript, this implies a summation over the index. In such expressions an index should not occur more than twice. For example, according to Einstein's summation convention, the summations in equation (1.62) and (1.63) can be expressed as

$$dx^{'\alpha} = \sum_{\beta=1}^{N} \frac{\partial x^{'\alpha}}{\partial x^{\beta}} dx^{\beta} = \Lambda_{\beta}^{\alpha}(x) dx^{\beta} \tag{1.64}$$

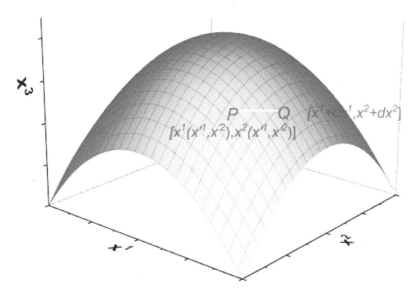

Figure 1.13. Point P and Q on a surface.

and

$$dx^\alpha = \sum_{\beta=1}^{N} \frac{\partial x^\alpha}{\partial x^{'\beta}} dx^{'\beta} = \Lambda_\beta^{'\alpha}(x')dx^{'\beta}, \tag{1.65}$$

where

$$\Lambda_\beta^\alpha(x) = \frac{\partial x^{'\alpha}}{\partial x^\beta}, \quad \Lambda_\beta^{'\alpha}(x') = \frac{\partial x^\alpha}{\partial x^{'\beta}}. \tag{1.66}$$

Note that the index α, which is known as the *free index*, can take any value from 1 to N. The index β, which is known as the *dummy index*, must be summed up from 1 to N.

1.4 The local geometry of a manifold

What we know about a manifold with dimension N is an amorphous collection of points with coordinates $(x^1, x^2, x^3 \cdots x^N)$ with no defined geometry. We can only visualize a two-dimensional manifold as it can be described in a three-dimensional Euclidean space. In figure 1.14 we see a collection of points that can be described by two coordinates (x^1, x^2) which display three different geometries. The points shaded green display a plane geometry, those shaded yellow display a cylindrical geometry, and those shaded red display a spherical geometry. We can define the geometry of a manifold locally. The local geometry of a manifold at point P with coordinates, for example for a two-dimensional manifold (x^1, x^2), is determined by defining the invariant 'distance' or interval, which we denote by ds, between point P and another point Q $(x^1 + dx^1, x^2 + dx^2)$ (see figure 1.15). To get from point P to point Q, suppose we traveled an infinitesimal distance, ds_1, on the curve defined by x^1 along

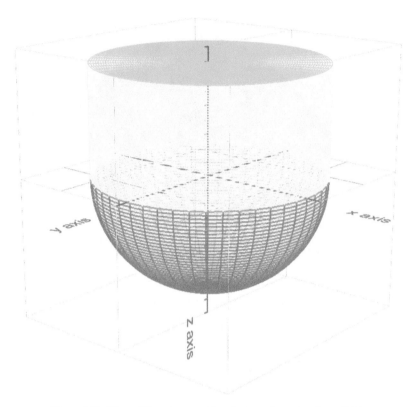

Figure 1.14. Three different geometries of a two-dimensional manifold.

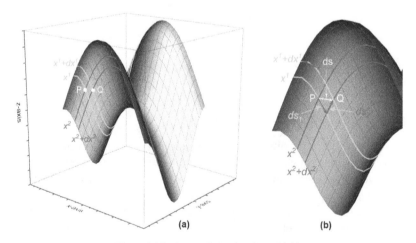

Figure 1.15. A two-dimensional manifold.

the direction shown by the first blue arrow and then we traveled a distance, ds_2, along the direction shown by the second blue arrow on the curve defined by x^2. Then the distance ds between points P and Q can be expressed as

$$(ds)^2 = (ds_1)^2 + (ds_2)^2 + 2(ds_1)(ds_2)\cos(\theta), \tag{1.67}$$

where θ is the angle between the first and second blue arrows. Suppose we represent the direction along the first blue arrow by the unit vector $\hat{e}_1$ and the direction along the second blue arrow by the unit vector $\hat{e}_2$, then one can write

$$d\vec{s}_1 = dx^1\hat{e}_1, \; d\vec{s}_2 = dx^2\hat{e}_2, \tag{1.68}$$

so that

$$(ds_1)^2 = \hat{e}_1 \cdot \hat{e}_1(dx^1)^2, \; (ds_2)^2 = \hat{e}_2 \cdot \hat{e}_2(dx^2)^2, \; \cos(\theta) = \hat{e}_1 \cdot \hat{e}_2. \tag{1.69}$$

Note that we are considering the general case where the two variables are non-orthogonal and the corresponding unit vectors not perpendicular. Upon substituting equations (1.68) and (1.69) into equation (1.67), we find

$$\begin{aligned}(ds)^2 &= \hat{e}_1 \cdot \hat{e}_1(dx^1)^2 + (\hat{e}_1 \cdot \hat{e}_2)dx^1dx^2 \\ &+ (\hat{e}_2 \cdot \hat{e}_1)dx^2dx^1 + \hat{e}_2 \cdot \hat{e}_2(dx^2)^2.\end{aligned} \tag{1.70}$$

The unit vectors $\hat{e}_1$ and $\hat{e}_2$, as one can see from the geometry of the surface shown in figure 1.15, depend on the curvature of the two paths. These paths are defined by x^1 and x^2 and the unit vectors $\hat{e}_1$ and $\hat{e}_2$ depend on (x^1, x^2). Taking this into account, one can then generally write equation (1.70) as

$$\begin{aligned}(ds)^2 &= g_{11}(x^1, x^2)dx^1dx^1 + g_{12}(x^1, x^2)dx^1dx^2 + g_{21}(x^1, x^2)dx^1dx^2 \\ &+ g_{22}(x^1, x^2)dx^2dx^2 = g_{\alpha\beta}(x)dx^\alpha dx^\beta,\end{aligned} \tag{1.71}$$

where we introduced the functions

$$\begin{aligned}g_{11}(x^1, x^2) &= \hat{e}_1(x^1, x^2) \cdot \hat{e}_1(x^1, x^2), \; g_{22}(x^1, x^2) = \hat{e}_2(x^1, x^2) \cdot \hat{e}_2(x^1, x^2) \\ g_{12}(x^1, x^2) &= g_{21}(x^1, x^2) = \hat{e}_1(x^1, x^2) \cdot \hat{e}_2(x^1, x^2),\end{aligned} \tag{1.72}$$

and used Einstein's summation convention. What we stated as unit vectors $\hat{e}_1$ and $\hat{e}_2$ will be discussed in more detail in the next chapter. For now, what we found in equation (1.71) is the equation that defines the local geometry of a two-dimensional manifold which we can visualize in a three-dimensional Euclidean space. Though it is impossible to visualize the geometry of an N-dimensional manifold for $N > 2$, the local geometry of an N-dimensional manifold is defined by the equation

$$(ds)^2 = g_{\alpha\beta}(x)dx^\alpha dx^\beta, \tag{1.73}$$

where $g_{\alpha\beta}(x) = g_{\alpha\beta}(x^1, x^2, \dots, x^N)$ is the component of the metric tensor field in our chosen coordinate system and $\alpha, \beta = 1, 2, 3 \dots N$. In the general theory of relativity, we are interested in a manifold where the geometry is described by the interval $(ds)^2$ in equation (1.73). The geometry of the manifold defined by equation (1.73) is known as *Riemannian geometry* if $(ds)^2 > 0$. As we will see in the Minkowski space-time manifold, the interval $(ds)^2$ can also be negative (space-like) or zero (light-like).

In such cases the geometry is referred to as *pseudo-Riemannian geometry* and the manifold can be referred to as *pseudo-Riemannian.* It is important to note that if the two curves defined by the coordinates x^1, x^2 are orthogonal at every point on the manifold, one can easily see that the angles between the two blue arrows shown in figure 1.15 become perpendicular to one another (i.e., $\hat{e}_1 \perp \hat{e}_2$), and we have

$$g_{12}(x^1, x^2) = g_{21}(x^1, x^2) = \hat{e}_1(x^1, x^2) \cdot \hat{e}_2(x^1, x^2) = 0, \tag{1.74}$$

and the interval becomes

$$(ds)^2 = g_{11}(x)dx^1dx^1 + g_{22}(x)dx^2dx^2 + \cdots + g_{NN}(x)dx^Ndx^N, \tag{1.75}$$

for *an* N-dimensional manifold. When

$$g_{\alpha\beta}(x) = \delta_{\alpha\beta},$$

we find

$$(ds)^2 = \delta_{\alpha\beta}dx^\alpha dx^\beta = (dx^1)^2 + (dx^2)^2 + \cdots + (dx^N)^2, \tag{1.76}$$

and such a metric is known as a *Euclidean metric.* The interval $(ds)^2$ is invariant under coordinate transformation, and sometimes it is much easier to do it in one coordinate system than another. In such cases it is important to know how to make the coordinate transformation. To this end, applying the relation in equation (1.65), we can express

$$dx^\alpha = \frac{\partial x^\alpha}{\partial x'^\kappa}dx'^\kappa, \ dx^\beta = \frac{\partial x^\beta}{\partial x'^d}dx'^d, \tag{1.77}$$

so that the interval can be transformed as

$$(ds)^2 = g_{\alpha\beta}(x)\frac{\partial x^\alpha}{\partial x'^\mu}\frac{\partial x^\beta}{\partial x'^\nu}dx'^\mu dx'^\nu = g'_{\kappa d}(x')dx'^\mu dx'^\nu, \tag{1.78}$$

where

$$g'_{\mu\nu}(x') = g_{\alpha\beta}(x)\frac{\partial x^\alpha}{\partial x'^\mu}\frac{\partial x^\beta}{\partial x'^\nu}. \tag{1.79}$$

Note that $g_{\alpha\beta}(x) = g_{\alpha\beta}(x^1, x^2, \ldots, x^N)$ and $g_{\alpha\beta}(x') = g_{\alpha\beta}(x'^1, x'^2, \ldots, x'^N)$.

Intrinsic and extrinsic geometry: the curvature in a given geometry with dimension M defined by the metric equation

$$(ds)^2 = g_{\alpha\beta}(x)dx^\alpha dx^\beta \tag{1.80}$$

is said to be *intrinsic* when the geometry remains unchanged when it is viewed in a higher dimension, N ($N > M$), otherwise it is is said to be *extrinsically curved.* A cylindrical surface is extrinsically curved as it possesses a geometry of a plane even when viewed in a three-dimensional space.

Example 1.3.
   (a) Find the interval for the plane shown in figure 1.16.
   (b) Show that the curvature of a cylindrical surface is *extrinsic*.

Solution:
   (a) Let us consider a point on a plane with coordinates (x, y, z) shown in figure 1.16.

   We recall that if $\vec{N} = a\hat{x} + b\hat{y} + c\hat{z}$ is normal to this plane, then the inner product of the vector $\vec{N}$ and the vector

$$\vec{r} - \vec{r_0} = (x - x_0)\hat{x} + (y - y_0)\hat{y} + (z - z_0)\hat{z} \qquad (1.81)$$

is zero:

$$\vec{N} \cdot (\vec{r} - \vec{r_0}) = a(x - x_0) + b(y - y_0) + c(z - z_0) = 0. \qquad (1.82)$$

This can be rewritten as

$$z(x, y) = -\frac{a}{c}x - \frac{b}{c}y + \frac{ax_0 + by_0 + cz_0}{c}. \qquad (1.83)$$

Introducing the constant

$$q = \frac{ax_0 + by_0 + cz_0}{c} \qquad (1.84)$$

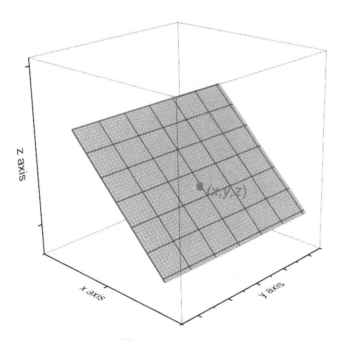

Figure 1.16. A plane.

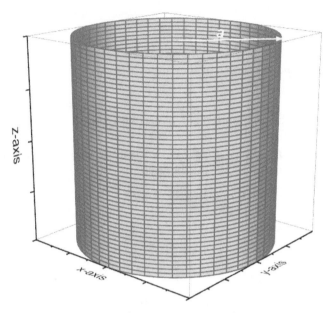

Figure 1.17. A surface with cylindrical geometry.

and the transformation of coordinates defined by

$$x^1 = -\frac{a}{c}x,\ x^2 = -\frac{b}{c}y,\ \Rightarrow z(x, y) = g(x^1, x^2),$$ (1.85)

the equation of a plane can be written:

$$f(x^1, x^2) = x^1 + x^2 + q.$$ (1.86)

Noting that

$$\hat{e}_1 = -\hat{x},\ \hat{e}_2 = -\hat{y},$$ (1.87)

we find

$$g_{11}(x^1, x^2) = \hat{e}_1(x^1, x^2) \cdot \hat{e}_1(x^1, x^2) = (-\hat{x}) \cdot (-\hat{x}) = 1,$$
$$g_{22}(x^1, x^2) = \hat{e}_2(x^1, x^2) \cdot \hat{e}_2(x^1, x^2) = (-\hat{y}) \cdot (-\hat{y}) = 1$$ (1.88)
$$g_{12}(x^1, x^2) = g_{21}(x^1, x^2) = 0,$$

so that the interval

$$(ds)^2 = g_{\alpha\beta}(x)dx^\alpha dx^\beta$$ (1.89)

for a plane becomes

$$(ds)^2 = (dx^1)^2 + (dx^2)^2.$$ (1.90)

(b) Now let us consider the cylindrical surface shown in figure 1.17, which we can construct by curving a two-dimensional plane. Suppose the cylinder has radius a. Using cylindrical coordinates a point on the surface of the cylinder can be described by (a, φ, z). We may write the interval between two points on this surface, point P with coordinates (a, φ, z) and point Q with coordinates $(a, \varphi + d\varphi, z + dz)$, as

$$(ds)^2 = (ad\varphi)^2 + (dz)^2. \tag{1.91}$$

This is a two-dimensional surface embedded in a three-dimensional Euclidean space. It has a cylindrically curved geometry when it is viewed in this three-dimensional Euclidean space. But it is possible to actually obtain the plane geometry in equation (1.90) by simply substituting

$$x^1 = a\varphi, \ x^2 = z \Rightarrow (ds)^2 = (dx^1)^2 + (dx^2)^2. \tag{1.92}$$

Note that such kind of geometry is *not intrinsic* and is instead called *extrinsic*: its curvature is extrinsic and is a result of the way it is embedded in the three-dimensional space.

Example 1.4. Show that the curvature of a spherical surface is intrinsic.

   Solution: In figure 1.18 we see a sphere where a point can be described by the coordinates (r, θ, φ). Let us assume the sphere has a radius $r = a$. Then the surface

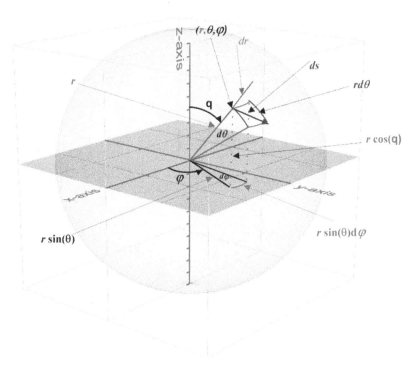

Figure 1.18. The interval ds for a two-dimensional sphere embedded in a three-dimensional Euclidean space. Adapted from [xx]. © IOP Publishing Ltd. All rights reserved.

defined by a pair of points on this sphere separated by a distance ds (see figure 1.18) can be expressed as

$$(ds)^2 = (ad\theta)^2 + (a \sin (\theta)d\varphi)^2 = a^2(d\theta)^2 + a^2 \sin^2 (\theta)(d\varphi)^2, \qquad (1.93)$$

or using the notation (x^2, x^3) for (θ, φ), we can write

$$(ds)^2 = (adx^2)^2 + a^2 \sin^2 (x^2)(dx^3)^2. \qquad (1.94)$$

This is a two-dimensional surface embedded in a three-dimensional Euclidean space. It is not possible to obtain this geometry from plane geometry like the two-dimensional cylindrical geometry. Such curved geometry is *intrinsic*. This means the geometry of a sphere is intrinsically curved because we cannot transform equation (1.94) to the Euclidean form,

$$(ds)^2 = (dx'^2)^2 + (dx'^3)^2, \qquad (1.95)$$

over the whole surface by any coordinate transformation. Note that this can be done locally but not for the whole spherical surface.

Example 1.5. Find the interval (metric) for a two-dimensional sphere of radius a embedded in a three-dimensional Euclidean space for the coordinates systems (x'^1, x'^2, x'^3) and (x^1, x^2, a) shown in figure 1.19. Note that the coordinate system

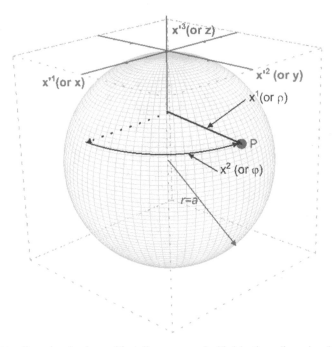

Figure 1.19. A two-dimensional sphere with radius $r = a$ embedded in three-dimensional Euclidean space described by two coordinate systems $(a, \rho, \varphi) \rightarrow (a, x^1, x^2)$ and $(x, y, z) \rightarrow (x'^1, x'^2, x'^3)$.

(x'^1, x'^2, x'^3) is the usual Cartesian coordinates with origin shifted by a along the positive z-axis:

$$x'^1 = x, \ x'^2 = y, \ x'^3 = z + a. \tag{1.96}$$

We recall that for a point on the surface of the sphere with the origin of the Cartesian coordinates at the origin $(0, 0, 0)$, we have

$$(x'^1)^2 + (x'^2)^2 + (x'^3)^2 = a^2, \tag{1.97}$$

so that for an origin at the north pole of the sphere $(0, 0, a)$, one can write

$$(x'^1)^2 + (x'^2)^2 + (x'^3 - a)^2 = a^2. \tag{1.98}$$

The coordinates (x'^1, x'^2, x'^3) and (x^1, x^2, a) are related by

$$x'^1 = x^1 \cos (x^2), \ x'^2 = x^1 \sin (x^2).$$

Solution: We recall that the line element, in a three-dimensional Euclidean space, is given by

$$(ds)^2 = (dx'^1)^2 + (dx'^2)^2 + (dx'^3)^2. \tag{1.99}$$

For the two-dimensional spherical geometry with radius a,

$$g(x'^1, x'^2, x'^3) = (x'^1)^2 + (x'^2)^2 + (x'^3 - a)^2 = a^2. \tag{1.100}$$

Upon taking the total differential, we find

$$dg(x'^1, x'^2, x'^3) = \frac{\partial g}{\partial x'^1}dx'^1 + \frac{\partial g}{\partial x'^2}dx'^2 + \frac{\partial g}{\partial x'^3}dx'^3 = 0$$

$$\Rightarrow 2x'^1 dx'^1 + 2x'^2 dx'^2 + 2(x'^3 - a)dx'^3 = 0 \tag{1.101}$$

$$\Rightarrow dx'^3 = -\frac{x'^1 dx'^1 + x'^2 dx'^2}{x'^3 - a}.$$

Noting that, from equation (1.100),

$$x'^3 - a = \sqrt{a^2 - (x'^1)^2 - (x'^2)^2}, \tag{1.102}$$

equation (1.101) can be rewritten as

$$dx'^3 = -\frac{x'^1 dx'^1 + x'^2 dx'^2}{\sqrt{a^2 - (x'^1)^2 - (x'^2)^2}}. \tag{1.103}$$

Then for a two-dimensional sphere embedded in a three-dimensional Euclidean space, the metric is given by

$$(ds)^2 = (dx'^1)^2 + (dx'^2)^2 + \frac{(x'^1 dx'^1 + x'^2 dx'^2)^2}{a^2 - (x'^1)^2 - (x'^2)^2}$$

$$= \{[(dx'^1)^2 + (dx'^2)^2][a^2 - (x'^1)^2 - (x'^2)^2] + (x'^1 dx'^1)^2 + (x'^2 dx'^2)^2$$

$$+ 2x'^1 x'^2 dx'^1 dx'^2\} / [a^2 - (x'^1)^2 - (x'^2)^2], \tag{1.104}$$

which simplifies into

$$(ds)^2 = \left\{ (a^2 - (x'^2)^2)(dx'^1)^2 + (a^2 - (x'^1)^2)(dx'^2)^2 \right.$$
$$\left. + (2x'^1 x'^2)dx'^1 dx'^2 \right\} / [a^2 - (x'^1)^2 - (x'^2)^2]. \tag{1.105}$$

This can be rewritten as

$$(ds)^2 = g_{\alpha\beta}(x')dx'^\alpha dx'^\beta, \tag{1.106}$$

for $\alpha, \beta = 1, 2$, where

$$g_{11} = [a^2 - (x'^2)^2] / [a^2 - (x'^1)^2 - (x'^2)^2],$$
$$g_{22} = [a^2 - (x'^1)^2] / [a^2 - (x'^1)^2 - (x'^2)^2], \tag{1.107}$$
$$g_{12} = g_{21} = x'^1 x'^2 / [a^2 - (x'^1)^2 - (x'^2)^2].$$

For a point in the neighborhood of the north pole of the sphere, we can make the approximation

$$x'^1 = x'^2 \simeq 0 \Rightarrow g_{11} = g_{22} = 1, \ g_{12} = g_{21} = 0, \tag{1.108}$$

and the metric in equation (1.105) reduces to the Euclidean form

$$(ds)^2 = (dx'^1)^2 + (dx'^2)^2. \tag{1.109}$$

Consider the coordinates (x^1, x^2) defined by the transformation

$$x'^1 = x^1 \cos(x^2), \ x'^2 = x^1 \sin(x^2)$$
$$\Rightarrow dx'^1 = dx^1 \cos(x^2) - x^1 \sin(x^2)dx^2, \tag{1.110}$$
$$dx'^2 = dx^1 \sin(x^2) + x^1 \cos(x^2)dx^2.$$

These coordinates along with the equation of a sphere,

$$(x'^1)^2 + (x'^2)^2 + (x'^3 - a)^2 = a^2, \tag{1.111}$$

define any point on the surface of the sphere. For any point on the surface of the sphere, in terms (x^1, x^2), one can write

$$x'^3 - a = \sqrt{a^2 - (x'^1)^2 - (x'^2)^2} = \sqrt{a^2 - (x^1 \cos(x^2))^2 - (x^1 \cos(x^2))^2}$$
$$\Rightarrow x'^3 - a = \sqrt{a^2 - (x^1)^2} \Rightarrow dx'^3 = -\frac{x^1 dx^1}{\sqrt{a^2 - (x^1)^2}}. \tag{1.112}$$

Then using equations (1.110) and (1.112), the interval in equation (1.99) can be rewritten as

$$(ds)^2 = (dx^1 \cos (x^2) - x^1 \sin (x^2) dx^2)^2$$
$$+ (dx^1 \sin (x^2) + x^1 \cos (x^2) dx^2)^2 + \left(\frac{x^1 dx^1}{\sqrt{a^2 - (x^1)^2}} \right)^2$$
$$= (dx^1)^2 + (x^1)^2 (dx^2)^2 + \frac{(x^1)^2 (dx^1)^2}{a^2 - (x^1)^2},$$
$$\Rightarrow (ds)^2 = \frac{a^2 (dx^1)^2}{a^2 - (x^1)^2} + (x^1)^2 (dx^2)^2. \tag{1.113}$$

Equation (1.113) can also be put in the form

$$(ds)^2 = g_{11}(dx^1)^2 + g_{22}(dx^2)^2 = g_{\mu\nu}(x) dx^\mu dx^\nu, \tag{1.114}$$

where $\mu, \nu = 1, 2$, and

$$g_{11} = \frac{a^2}{a^2 - (x^1)^2}, \; g_{22} = (x^1)^2, \; g_{12} = g_{21} = 0 \tag{1.115}$$

are the elements of the metric tensor. We will see the use of these elements of the metric tensor in the next section to determine the length and area of a two-dimensional sphere embedded in a three-dimensional Euclidean space.

Example 1.6. Determine the metric for a three-dimensional sphere of radius a embedded in a four-dimensional Euclidean space. This three-dimensional sphere is defined by the constraint

$$g(x'^1, x'^2, x'^3, x'^4) = (x'^1)^2 + (x'^2)^2 + (x'^3)^2 + (x'^4)^2 = a^2. \tag{1.116}$$

Solution: We can write the interval

$$(ds)^2 = (dx'^1)^2 + (dx'^2)^2 + (dx'^3)^2 + (dx'^4)^2. \tag{1.117}$$

For a three-dimensional spherical geometry with radius a, from equation (1.116) we have

$$x'^4 = \sqrt{a^2 - (x'^1)^2 - (x'^2)^2 - (x'^3)^2} \tag{1.118}$$

so that

$$dx'^4 = \frac{\partial x'^4}{\partial x'^1} dx'^1 + \frac{\partial x'^4}{\partial x'^2} dx'^2 + \frac{\partial x'^4}{\partial x'^3} dx'^3$$
$$= -[x'^1 dx'^1 + x'^2 dx'^2 + x'^3 dx'^3]/x'^4, \tag{1.119}$$

which leads to

$$(dx'^4)^2 = \left\{ (x'^1)^2 (dx'^1)^2 + (x'^2)^2 (dx'^2)^2 + (x'^3)^2 (dx'^3)^2 + 2x'^1 x'^2 dx'^1 dx'^2 \right.$$
$$\left. + 2x'^1 x'^3 dx'^1 dx'^3 + 2x'^2 x'^3 dx'^2 dx'^3 \right\} \Big/ x'^4. \tag{1.120}$$

Then substituting equation (1.120) into the metric in equation (1.117), we find

$$(ds)^2 = \left\{ [a^2 - (x'^2)^2 - (x'^3)^2](dx'^1)^2 + [a^2 - (x'^1)^2 - (x'^3)^2](dx'^2)^2 \right.$$
$$+ [a^2 - (x'^1)^2 - (x'^2)^2](dx'^3)^2 + 2x'^1 x'^2 dx'^1 dx'^2 + 2x'^1 x'^3 dx'^1 dx'^3 \tag{1.121}$$
$$\left. + 2x'^2 x'^3 dx'^2 dx'^3 \right\} /(x'^4)^2.$$

Equation (1.121) can be rewritten as

$$(ds)^2 = g'_{11}(dx'^1)^2 + g'_{22}(dx'^2)^2 + g'_{33}(dx'^3)^2 + g'_{12}dx'^1 dx'^2 + g'_{21}dx'^2 dx'^1$$
$$+ g'_{13}dx'^1 dx'^3 + g'_{31}dx'^3 dx'^1 + g'_{23}dx'^2 dx'^3 + g'_{32}dx'^3 dx'^2 \tag{1.122}$$
$$\Rightarrow (ds)^2 = g'_{\alpha\beta}(x')dx'^\alpha dx'^\beta,$$

where $\alpha, \beta = 1, 2, 3$, and

$$g_{11}(x') = [a^2 - (x'^2)^2 - (x'^3)^2]/(x'^4)^2,$$
$$g'_{22}(x') = [a^2 - (x'^1)^2 - (x'^3)^2]/(x'^4)^2, \tag{1.123}$$
$$g'_{33}(x') = [a^2 - (x'^1)^2 - (x'^2)^2]/(x'^4)^2,$$

$$g'_{12}(x') = g'_{21}(x') = 2x'^1 x'^2 / x'^4, \quad g'_{13}(x') = g'_{31}(x') = 2x'^1 x'^3 /(x'^4)^2,$$
$$g'_{23}(x') = g'_{32}(x') = 2x'^2 x'^3 /(x'^4)^2 \tag{1.124}$$

are the elements for the metric tensor. Note that x'^4, given by equation (1.118), is a function of the coordinates (x'^1, x'^2, x'^3). Next, we determine the metric in terms of another set of coordinates (r, θ, φ), which we represent by (x^1, x^2, x^3), and related to the coordinates (x'^1, x'^2, x'^3) by the transformation

$$x = x'^1 = r \sin(\theta) \cos(\varphi), \; y = x'^2 = r \sin(\theta) \sin(\varphi), \; z = x'^3 = r \cos(\theta), \tag{1.125}$$

or

$$x'^1 = x^1 \sin(x^2) \cos(x^3), \; x'^2 = x^1 \sin(x^2) \sin(x^3), \; x'^3 = x^1 \cos(x^2). \tag{1.126}$$

In figure 1.20(a) we see part of a sphere with radius r. A point on the surface of this sphere is described by the coordinates (r, θ, φ) or (x^1, x^2, x^3). From this point (P) we move on this sphere along the θ for an infinitesimal displacement $rd\theta$, along the φ direction another displacement $r \sin(\theta)d\varphi$, and we jump off the surface of the sphere along the radial direction for a displacement of dr to reach point Q as shown in figure 1.20(b). Note that the x-axis, y-axis, and z-axis correspond to (x'^1, x'^2, x'^3). The 'distance' squared between point P and Q can easily be determined using the Pythagorean theorem (see figure 1.21(a) and (b)). First find the length of the hypotenuse ds_s (yellow) for a triangle on the surface of the blue sphere (radius r):

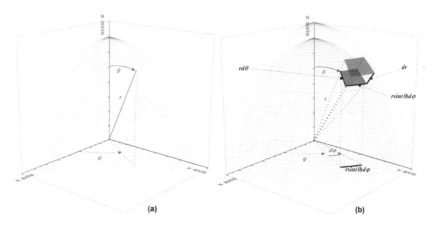

Figure 1.20. (a) A point on a sphere with coordinates (r, θ, φ) and (b) in infinitesimal displacements from this point along θ, φ, and r.

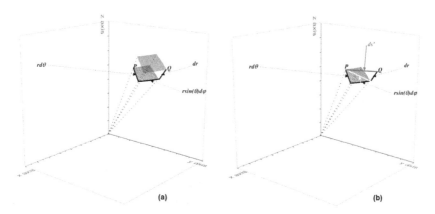

Figure 1.21. A closer look at the infinitesimal displacements.

$$(ds_s)^2 = (r \sin (\theta) d\varphi)^2 + (r d\theta)^2 = (x^1 \sin (x^2) dx^3)^2 + (x^1 dx^2)^2. \qquad (1.127)$$

Then the distance between P and Q, shown by the red line, can be expressed as

$$\begin{aligned}
\left(ds_{xyz}\right)^2 &= (dr)^2 + (ds_s)^2 = (dr)^2 + r^2 \sin^2 (\theta)(d\varphi)^2 + r^2 (d\theta)^2 \\
&= (dx^1)^2 + (x^1 \sin (x^2) dx^3)^2 + (x^1 dx^2)^2.
\end{aligned} \qquad (1.128)$$

Note that we refer to this distance as ds_{xyz} because it represents a distance

$$\left(ds_{xyz}\right)^2 = dx^2 + dy^2 + dz^2 = (dx^{'1})^2 + (dx^{'2})^2 + (dx^{'3})^2. \qquad (1.129)$$

We are considering a three-dimensional sphere embedded in a four-dimensional Euclidean space, where the interval between point P and Q is given by

$$(ds)^2 = \left(ds_{xyz}\right)^2 + (dx^{'4})^2 = (dx^{'1})^2 + (dx^{'2})^2 + (dx^{'3})^2 + (dx^{'4})^2, \qquad (1.130)$$

with the constraint

$$x'^4 = \sqrt{a^2 - (x'^1)^2 - (x'^2)^2 - (x'^3)^2}, \tag{1.131}$$

which can be rewritten, in terms of the coordinates (x^1, x^2, x^3) in equation (1.126), as

$$x'^4 = \sqrt{a^2 - (x^1)^2} \Rightarrow dx'^4 = -\frac{x^1 dx^1}{\sqrt{a^2 - (x^1)^2}}. \tag{1.132}$$

Then the metric for a three-dimensional sphere embedded in a four-dimensional Euclidean space is given by

$$(ds)^2 = \left(ds_{xyz}\right)^2 + (dx'^4)^2 = dr^2 + r^2 \sin^2(\theta)d\varphi^2 + r^2 d\theta^2 + \frac{r^2 dr^2}{a^2 - r^2}$$

$$= \frac{a^2}{a^2 - r^2}dr^2 + r^2 d\theta^2 + r^2 \sin^2(\theta)d\varphi^2 \tag{1.133}$$

$$= \frac{a^2 (dx^1)^2}{a^2 - (x^1)^2} + (x^1 \sin(x^2)dx^3)^2 + (x^1 dx^2)^2. \tag{1.134}$$

Equation (1.134) can be put in the form

$$(ds)^2 = g_{11}(dx^1)^2 + g_{22}(dx^2)^2 + g_{33}(dx^3)^2 = g_{\mu\nu}(x)dx^\mu dx^\nu, \tag{1.135}$$

where

$$g_{11}(x) = \frac{a^2}{a^2 - (x^1)^2}, \ g_{22}(x) = (x^1)^2, \ g_{33}(x) = (x^1)^2 \sin^2(x^2),$$

$$g_{12}(x) = g_{21}(x) = g_{13}(x) = g_{31}(x) = g_{23}(x) = g_{32}(x) = 0, \tag{1.136}$$

and $\mu, \nu = 1, 2, 3$. We will see the use of these elements of the metric tensor in the next section, where we determine the length, area, and volume of a three-dimensional sphere embedded in a four-dimensional Euclidean space.

1.5 Length, area, and volume

Length
Consider two points, P and Q, on a manifold of dimension N. Suppose these two points are connected by some curve. For example, for a two-dimensional manifold embedded in a three-dimensional Euclidean space shown in figure 1.22, P and Q are connected by the blue curve. The length of the curve connecting these two points is given by

$$L_{PQ} = \int_P^Q \sqrt{|(ds)^2|} = \int_P^Q \sqrt{|g_{\mu\nu}(x)dx^\mu dx^\nu|}. \tag{1.137}$$

If the coordinates in the manifold are parameterized by u, $x^\alpha = x^\alpha(u)$, we have

$$dx^\mu = \frac{dx^\mu}{du}du, \ dx^\nu = \frac{dx^\nu}{du}du, \tag{1.138}$$

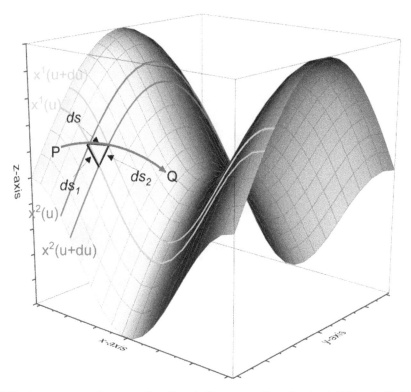

Figure 1.22. A curve connecting two points (P and Q) on a two-dimensional manifold embedded in three-dimensional Euclidean space.

and the length can be expressed as

$$L_{PQ} = \int_P^Q \sqrt{\left| g_{\mu\nu}(u)\frac{dx^\mu}{du}\frac{dx^\nu}{du} \right|}\, du. \qquad (1.139)$$

Note that the absolute value for the interval $| (ds)^2 |$ is because of the fact that for pseudo-Riemannian manifolds the interval can be negative, as is the case for the "spacelike" four-dimensional Minkowski manifold.

Area and volume

A closer look at an infinitesimal area (the pink region) on a two-dimensional manifold embedded in a three-dimensional Euclidean space is shown in figure 1.23. In this figure, we can see that the curves defined by the coordinates x^1 and x^2 are intersecting at 90°. Such coordinate systems are orthogonal. For an orthogonal coordinate system, the infinitesimal area can easily be written as

$$dA = | \vec{ds}_1 \times \vec{ds}_2 |. \qquad (1.140)$$

For an orthogonal coordinate system, the metric (see figure 1.22)

$$(ds)^2 = (ds_1)^2 + (ds_2)^2 \qquad (1.141)$$

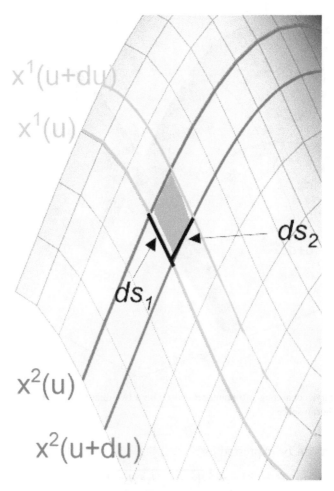

Figure 1.23. An infinitesimal area on a two-dimensional manifold embedded in three-dimensional Euclidean space.

is diagonal,

$$(ds)^2 = \left(\sqrt{g_{11}}\,dx^1\right)^2 + \left(\sqrt{g_{22}}\,dx^2\right)^2 \cdots + \left(\sqrt{g_{NN}}\,dx^N\right)^2$$
$$\Rightarrow ds_1 = \sqrt{g_{11}}\,dx^1,\ ds_2 = \sqrt{g_{22}}\,dx^2,$$

(1.142)

and the area of the surface defined by

$$x^\alpha = \text{constant},$$

(1.143)

for, $\alpha = 3, 4 \ldots N$, becomes

$$dA = |\,d\vec{s}_1 \times d\vec{s}_2\,| = \sqrt{|\,g_{11}(x)g_{22}(x)\,|}\,dx^1 dx^2.$$

(1.144)

In the case of an infinitesimal three-volume inside a surface that depends on three coordinates (x^1, x^2, x^3) and is defined by

$$x^\alpha = \text{constant} \qquad (1.145)$$

for $\alpha = 4, 5...N$, the infinitesimal volume is given by

$$d^3V = \sqrt{|\, g_{11}(x)g_{22}(x)g_{33}(x)\,|}\, dx^1 dx^2 dx^3. \qquad (1.146)$$

In the case of an infinitesimal four-volume inside a surface that depends on three coordinates (x^1, x^2, x^3, x^3) and is defined by

$$x^\alpha = \text{constant} \qquad (1.147)$$

for $\alpha = 5...N$, the infinitesimal volume is given by

$$d^4V = \sqrt{|\, g_{11}(x)g_{22}(x)g_{33}(x)g_{44}(x)\,|}\, dx^1 dx^2 dx^3. \qquad (1.148)$$

This can be generalized to

$$d^N V = \sqrt{|\, g_{11}(x)g_{22}(x)\cdots g_{NN}(x)\,|}\, dx^1 dx^2 \cdots dx^N, \qquad (1.149)$$

for an infinitesimal N-volume inside the surface that depends on N coordinates $(x^1, x^2, ...x^N)$.

Example 1.7. For a two-dimensional sphere of radius a embedded in a three-dimensional Euclidean space consider part of the region from the north pole up to a circle defined by a radius $\rho = R$ ($x^1 = R$) shown in figure 1.24. Note that the origin of both coordinate systems is at the north pole of the sphere (see example 1.5).
  (a) Find the distance D from the origin to the perimeter of this circle along a line of constant φ (i. e. $x^2 = $ const.).
  (b) Find the circumference of the circle with radius $\rho = R$ (i. e. $x^1 = R = $ const.).
  (c) Find the area of the part of the spherical surface from the north pole up to the circle with radius $\rho = R$, (i.e., the surface shaded cyan in figure 1.24).

Solution:
  (a) We recall from example 1.5 that the metric for a two-dimensional sphere in a three-dimensional Euclidean manifold is given by

$$(ds)^2 = g_{11}(x)(dx^1)^2 + g_{22}(x)(dx^2)^2, \qquad (1.150)$$

where we used the notation (x^1, x^2) for the coordinates (ρ, φ) and the elements of the metric tensor

$$g_{11}(x) = \frac{a^2}{a^2 - (x^1)^2}, \; g_{22}(x) = (x^1)^2, \; g_{12}(x) = g_{21}(x) = 0. \qquad (1.151)$$

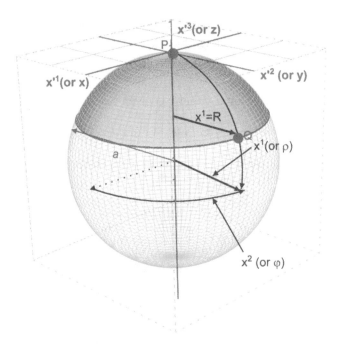

Figure 1.24. A two-dimensional sphere with radius a embedded in three-dimensional Euclidean space described by two coordinate systems (x'^1, x'^2) and $(x^1, x^2 \to \rho, \varphi)$.

We recall the length of a curve between two points P and Q on a manifold in terms of the metric tensor is given by

$$L_{PQ} = \int_P^Q \sqrt{\left| g_{\mu\nu}(x)dx^\mu dx^\nu \right|} . \tag{1.152}$$

For the curve shown in brown in figure 1.24 the coordinates for point P are $(x^1 = \rho = 0, x^2 = \varphi = \text{constant})$ and for point Q

$$x^1 = \rho = R, \quad x^2 = \varphi = \text{constant}, \tag{1.153}$$

which leads to

$$dx^2 = d\varphi = 0$$
$$\Rightarrow g_{12}(x)dx^1 dx^2 = g_{21}(x)dx^2 dx^1 = g_{22}(x)dx^2 dx^2 = 0. \tag{1.154}$$

Then the length becomes

$$D = \int_P^Q \sqrt{\left| g_{11}(x)dx^1 dx^1 \right|} = \int_0^R \frac{a}{\sqrt{a^2 - (x^1)^2}} dx^1 = a \sin^{-1}\left[\frac{R}{a}\right]. \tag{1.155}$$

(b) Along the circumference (the curve shown in pink in figure 1.24), we know that

$$x^1 = \rho = R = \text{constant}, \tag{1.156}$$

which leads to

$$dx^1 = d\rho = 0$$
$$\Rightarrow g_{12}(x)dx^1dx^2 = g_{21}(x)dx^2dx^1 = g_{11}(x)dx^2dx^2 = 0, \tag{1.157}$$
$$g_{22}(x) = (x^1)^2 \mid_{\rho=R} = R^2.$$

Thus the expression

$$L_{PQ} = \int_P^Q \sqrt{\mid g_{\mu\nu}(x)dx^\mu dx^\nu \mid} \tag{1.158}$$

for the circumference becomes

$$C = \int_0^{2\pi} \sqrt{\mid g_{22}(x)(dx^2)^2 \mid} = \int_0^{2\pi} Rdx^2 = \int_0^{2\pi} Rd\varphi = 2\pi R. \tag{1.159}$$

(c) According to equation (1.144), the infinitesimal area is given by

$$dA = \sqrt{\mid g_{11}(x)g_{22}(x) \mid}\, dx^1 dx^2. \tag{1.160}$$

Noting that

$$\sqrt{\mid g_{11}(x)g_{22}(x) \mid} = \sqrt{\left| \frac{a^2(x^1)^2}{a^2 - (x^1)^2} \right|} = \frac{ax^1}{\sqrt{a^2 - (x^1)^2}} = \frac{a}{\sqrt{a^2 - \rho^2}}, \tag{1.161}$$

then the surface area becomes

$$A = \int\int \sqrt{\mid g_{11}(x)g_{22}(x) \mid}\, dx^1 dx^2 = \int\int \frac{ax^1 dx^1 dx^2}{\sqrt{a^2 - (x^1)^2}}. \tag{1.162}$$

To cover the area (colored cyan) shown in figure 1.24, we should have for the limits of integrations $[0, R]$ for ρ and $[0, 2\pi]$ for φ, and the area becomes

$$A = a \int_0^R \frac{x^1 dx^1}{\sqrt{a^2 - (x^1)^2}} \int_0^{2\pi} dx^2 = 2\pi a^2 \left(1 - \sqrt{1 - \frac{R^2}{a^2}}\right). \tag{1.163}$$

Example 1.8. In example 1.6 we have determined the metric for a three-dimensional sphere with radius a embedded in a four-dimensional Euclidean space. Inside this three-dimensional sphere, let us consider a two-dimensional sphere with radius $r = R$.
  (a) Find the distance from its center to the surface of this two-dimensional sphere along constant θ ($x^2 = $ cons.) and constant φ ($x^3 = $ cons.).
  (b) Find the circumference across the equator of this two-dimensional sphere.
  (c) Find the total area of this two-dimensional spherical surface.
  (d) Find the volume bounded by this two-dimensional spherical surface.

Solution:

(a) We recall from example 1.6 that the metric for a three-dimensional sphere embedded in a four-dimensional Euclidean space is given by

$$(ds)^2 = g_{11}(dx^1)^2 + g_{22}(dx^2)^2 + g_{33}(dx^3)^2, \tag{1.164}$$

where we used the notation (x^1, x^2, x^3) for the coordinates (r, θ, φ) and

$$g_{11}(x) = \frac{a^2}{a^2 - (x^1)^2}, \quad g_{22}(x) = (x^1)^2, \quad g_{33}(x) = (x^1)^2 \sin^2(x^2),$$
$$g_{12}(x) = g_{21}(x) = g_{13}(x) = g_{31}(x) = g_{23}(x) = g_{32}(x) = 0. \tag{1.165}$$

For $x^2 = \theta = $ constant and $x^3 = \varphi = $ constant, we have

$$dx^2 = dx^3 = 0,$$
$$\Rightarrow g_{22}(x)dx^2dx^2 = g_{33}(x)dx^3dx^3 = g_{12}(x)dx^1dx^2 \tag{1.166}$$
$$= g_{23}(x)dx^2dx^3 = g_{13}(x)dx^1dx^3 = 0,$$

so that the distance D,

$$L_{PQ} = \int_P^Q \sqrt{\left| g_{\mu\nu}(x)dx^\mu dx^\nu \right|}, \tag{1.167}$$

becomes

$$L_{PQ} = \int_P^Q \sqrt{\left| g_{11}(x)dx^1dx^1 \right|} = \int_0^R \sqrt{\left| \frac{a^2}{a^2 - (x^1)^2}(dx^1)^2 \right|}$$
$$= \int_0^R \frac{a}{\sqrt{a^2 - (x^1)^2}} dx^1 \Rightarrow L_{PQ} = a \sin^{-1}\left[\frac{R}{a} \right], \tag{1.168}$$

which is the same as the result we obtained in the previous example.

(b) Across the equator, we have

$$x^1 = R, \quad x^2 = \pi/2 \Rightarrow dx^1 = dx^2 = 0, \tag{1.169}$$

so that

$$L_{PQ} = \int_P^Q \sqrt{\left| g_{\alpha\beta}(x)dx^\alpha dx^\beta \right|}, \tag{1.170}$$

for the circumference, C, it becomes

$$C = \int \sqrt{\left| g_{33}(x)(dx^3)^2 \right|}. \tag{1.171}$$

Noting that for $x^1 = R$ and $x^2 = \pi/2$ the metric element $g_{33}(x)$ becomes

$$g_{33}(x) = (x^1)^2 \sin^2(x^2) = R^2, \tag{1.172}$$

one finds for the circumference in equation (1.171):

$$C = \int_0^{2\pi} R dx^3 = 2\pi R.$$ (1.173)

(c) The two-dimensional sphere is defined by $x^1 = R =$ constant. Therefore, in view of equation (1.144), the infinitesimal area on this surface should be expressed as

$$dA = \sqrt{|g_{22}(x)g_{33}(x)|}\, dx^2 dx^3.$$ (1.174)

Noting that for $x^1 = R$,

$$g_{22}(x) = (x^1)^2\,|_{r=R} = R^2, \quad g_{33}(x) = (x^1)^2 \sin^2(x^2)\,|_{r=R} = R^2 \sin^2(x^2),$$ (1.175)

and the limit of integrations for $x^2 = \theta$ is $(0, \pi)$ and for $x^3 = \varphi$ is $(0, 2\pi)$, one finds for the surface area

$$A = \int_0^\pi \int_0^{2\pi} \sqrt{|R^4 \sin^2(x^2)|}\, dx^2 dx^3 = 4\pi R^2,$$ (1.176)

as expected.

(d) We are here considering a three-dimensional sphere embedded in a four-dimensional Euclidean space. As such, the volume is defined by $x^4 =$ constant. Therefore, applying the relation in equation (1.146), an infinitesimal volume in this four-dimensional space is given by

$$dV = \sqrt{|g_{11}(x)g_{22}(x)g_{33}(x)|}\, dx^1 dx^2 dx^3.$$ (1.177)

Noting that

$$g_{11}(x) = \frac{a^2}{a^2 - (x^1)^2}, \quad g_{22}(x) = (x^1)^2, \quad g_{33}(x) = (x^1)^2 \sin^2(x^2)$$

$$\Rightarrow g_{11}(x)g_{22}(x)g_{33}(x) = \frac{a^2(x^1)^4 \sin^2(x^2)}{a^2 - (x^1)^2},$$ (1.178)

for the volume bounded by the two-dimensional spherical surface of radius R, we have

$$V = \int_0^R \int_0^\pi \int_0^{2\pi} \sqrt{\frac{a^2(x^1)^4 \sin^2(x^2)}{a^2 - (x^1)^2}}\, dx^1 dx^2 dx^3$$

$$= \int_0^R \int_0^\pi \int_0^{2\pi} \frac{a(x^1)^2 \sin(x^2)}{\sqrt{a^2 - (x^1)^2}}\, dx^1 dx^2 dx^3$$ (1.179)

$$= a \int_0^R \frac{(x^1)^2 dx^1}{\sqrt{a^2 - (x^1)^2}} \int_0^\pi \sin(x^2) dx^2 \int_0^{2\pi} dx^3 = 4\pi a \int_0^R \frac{(x^1)^2 dx^1}{\sqrt{a^2 - (x^1)^2}},$$

where we replaced

$$\int_0^\pi \sin(x^2)dx^2 \int_0^{2\pi} dx^3 = 4\pi.$$

Let us evaluate the integral

$$I = \int_0^R \frac{(x^1)^2 dx^1}{\sqrt{a^2 - (x^1)^2}}. \tag{1.180}$$

Introducing the transformation defined by

$$x^1 = a\sin(v) \Rightarrow r = a\sin(v) \Rightarrow dx^1 = a\cos(v)dv, \ \sqrt{a^2 - (x^1)^2} = \cos(v), \tag{1.181}$$

one can rewrite the integral as

$$I = a^2 \int \sin^2(v)dv = \frac{a^2}{2}[v - \sin(v)\cos(v)]. \tag{1.182}$$

Noting that

$$x^1 = a\sin(v) = a\sin(v) \Rightarrow \sin(v) = \begin{cases} \dfrac{R}{a}, & \text{for } r = R, \\ 0, & \text{for } r = 0, \end{cases}$$

$$\Rightarrow \cos(v) = \sqrt{1 - \sin^2(v)} = \begin{cases} \sqrt{1 - \left(\dfrac{R}{a}\right)^2}, & \text{for } r = R, \\ 0, & \text{for } r = 0, \end{cases} \tag{1.183}$$

one finds for the volume

$$V = 4\pi a \int_0^R \frac{(x^1)^2 dx^1}{\sqrt{a^2 - (x^1)^2}} = 2\pi a^3 [v - \sin(v)\cos(v)] \Big|_0^{\sqrt{1-\left(\frac{R}{a}\right)^2}}$$

$$\Rightarrow V = 2\pi a^3 \left\{ \sin^{-1}\left(\frac{R}{a}\right) - \frac{R}{a}\sqrt{1 - \left(\frac{R}{a}\right)^2} \right\}. \tag{1.184}$$

One must be able to recover the three-dimensional Euclidean space for $a \to \infty$. This means that for $\frac{R}{a} \ll 1$, the result for the volume of a sphere with radius R must be that of the volume of a sphere with radius R in three-dimensional Euclidean space:

$$V = \frac{4}{3}\pi R^3. \tag{1.185}$$

Using a *Taylor series* expansion, one can make the approximations

$$\sqrt{1 - \left(\frac{R}{a}\right)^2} \simeq 1 - \frac{1}{2}\left(\frac{R}{a}\right)^2,$$

$$\sin^{-1}\left(\frac{R}{a}\right) \simeq \frac{R}{a} + \frac{1}{6}\left(\frac{R}{a}\right)^3,$$

(1.186)

for $\frac{R}{a} \ll 1$. Substituting these results into equation (1.184), we find

$$V = 2\pi a^3 \left\{ \frac{R}{a} + \frac{1}{6}\left(\frac{R}{a}\right)^3 - \frac{R}{a}\left(1 - \frac{1}{2}\left(\frac{R}{a}\right)^2\right) \right\}$$

$$= 2\pi a^3 \left\{ \frac{1}{6}\left(\frac{R}{a}\right)^3 + \frac{1}{2}\left(\frac{R}{a}\right)^3 \right\} \Rightarrow V = \frac{4}{3}\pi R^3.$$

(1.187)

1.6 Local Cartesian coordinates and tangent space

Generally, $(ds)^2$ can be positive, negative, or zero, as we saw in pseudo-Riemannian spaces. For now, we shall consider only Riemannian space where the metric

$$(ds)^2 = g_{\mu\nu}(x)dx^\mu dx^\nu \tag{1.188}$$

is positive. Even for Riemannian geometry it is not possible, in general, to find a coordinate transformation that converts the metric into a Euclidean form,

$$(ds)^2 = g_{\mu\nu}(x)dx^\mu dx^\nu = \delta_{\alpha\beta}dx'^\alpha dx'^\beta$$

$$= (dx'^1)^2 + (dx'^2)^2 + \cdots + (dx'^N)^2,$$

(1.189)

for all points on the manifold. This is because the coordinate transformation for an N-dimensional manifold provides only N equations $x^\alpha(x)$, but there are $N(N + 1)/2$ elements to be determined for the metric tensor $g'_{\alpha\beta}(x')$. For example, for a two-dimensional spherical manifold $N = 2$, we have two equations of transformation:

$$x'^1 = x^1 \cos(x^2), \ x'^2 = x^1 \sin(x^2), \tag{1.190}$$

but we need to find $N(N + 1)/2 = 3$ functions:

$$g'_{11}(x'^1, x'^2), \ g'_{22}(x'^1, x'^2), \ \text{and} \ g'_{12}(x'^1, x'^2) = g'_{21}(x'^1, x'^2). \tag{1.191}$$

However, it is possible to find local Cartesian coordinates x'^α at the point P on the manifold such that

$$(ds)^2 = \delta_{\alpha\beta}dx'^\alpha dx'^\beta = (dx'^1)^2 + (dx'^2)^2 + \cdots + (dx'^N)^2. \tag{1.192}$$

Before we state the condition for the metric to get such a coordinate transformation, let us consider a series expansion for a function of one variable, $f(x)$, shown in

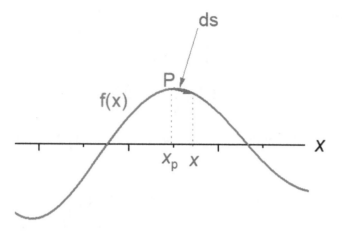

Figure 1.25. A function of a single variable.

figure 1.25. We recall that for any function $f(x)$ that is *differentiable for all values of x in the specified domain*,

$$\frac{d^n f(x)}{dx^n} \text{ exists for all } n \geqslant 0 , \qquad (1.193)$$

one can then write the *Taylor series* expansion about $x = x_p$ as

$$f(x) = f(x_p) + \frac{df(x)}{dx}\bigg|_{x=x_p} (x - x_p) + \frac{1}{2!}\frac{d^2 f(x)}{dx^2}\bigg|_{x=x_p} (x - x_p)^2 + \cdots. \qquad (1.194)$$

In the neighborhood of point P, where $|x - x_p| \ll 1$, we can make the approximation

$$f(x) \simeq f(x_p) + \frac{df(x)}{dx}\bigg|_{x=x_p} (x - x_p). \qquad (1.195)$$

We note that

$$m = \frac{df(x)}{dx}\bigg|_{x=x_p} \qquad (1.196)$$

gives the slope of the tangent line at point P. If this tangent line is horizontal (parallel to the x-axis), we find

$$m = \frac{df(x)}{dx}\bigg|_{x=x_p} = 0 \Rightarrow f(x) \simeq f(x_p) \Rightarrow ds = dx. \qquad (1.197)$$

Now let us consider a surface defined by the function of two coordinates $f(x) = f(x'^1, x'^2)$ shown in figure 1.26. In this case, for the sake of generality the coordinates x'^1 and x'^2 are curved coordinates. The series expansion in equation (1.194) becomes

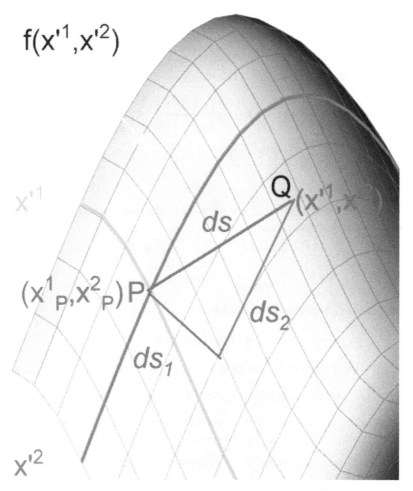

Figure 1.26. A surface defined by the function $f(x'^1, x'^2)$.

$$x'^1, x'^2) = f\left(x_p^1, x_p^2\right) + \frac{\partial f(x'^1, x'^2)}{\partial x'^1}\bigg|_{x'=x_p^1, x_p^2} (x'^1 - x_p^1)$$

$$+ \frac{\partial f(x'^1, x'^2)}{\partial x'^2}\bigg|_{x'=x_p^1, x_p^2} (x'^2 - x_p^2) + \frac{1}{2!}\left\{\frac{\partial^2 f(x'^1, x'^2)}{\partial (x'^1)^2}\bigg|_{x'=x_p^1, x_p^2} (x'^1 - x_p^1)^2\right.$$

$$+ 2\frac{\partial^2 f(x'^1, x''^2)}{\partial x'^1 \partial x'^1}\bigg|_{x'=x_p^1, x_p^2} (x'^1 - x_p^1)(x'^2 - x_p^2)$$

$$+ \left.\frac{\partial^2 f(x'^1, x''^2)}{\partial (x'^2)^2}\bigg|_{x'=x_p^1, x_p^2} (x'^2 - x_p^2)^2\right\} + \cdots$$

$$(1.198)$$

Note that from this expression it is easy to imagine how complex it gets for a function of N variables, like the metric $f(x') = g_{\alpha\beta}(x') = g_{\alpha\beta}(x'^1, x'^2, x'^3, \ldots x'^N)$.

In the neighborhood of point P where $\left| x'^1 - x_p^1 \right|$ and $\left| x'^2 - x_p^2 \right| \ll 1$, in a similar manner one can make the approximation

$$f(x'^1, x'^2) \simeq f\left(x_p^1, x_p^2\right) + \frac{\partial f(x'^1, x'^2)}{\partial x'^1}\bigg|_{x'=x_p^1, x_p^2} (x'^1 - x_p^1)$$

$$+ \frac{\partial f(x'^1, x'^2)}{\partial x'^2}\bigg|_{x'=x_p^1, x_p^2} (x'^2 - x_p^2). \tag{1.199}$$

When the tangent lines at point P are parallel to the two curved axes defined by the coordinates x'^1 and x'^2, we have

$$\frac{\partial f(x'^1, x'^2)}{\partial x'^1}\bigg|_{x'=x_p^1, x_p^2} = \frac{\partial f(x'^1, x'^2)}{\partial x'^2}\bigg|_{x'=x_p^1, x_p^2} = 0, \tag{1.200}$$

and the function can be approximated as

$$f(x'^1, x'^2) \simeq f\left(x_p^1, x_p^2\right). \tag{1.201}$$

Therefore, in the neighborhood of point P, one can then write

$$(ds)^2 = (ds_1)^2 + (ds_2)^2 \simeq (dx'^1)^2 + (dx'^1)^2. \tag{1.202}$$

Such coordinates are called *local Cartesian coordinates* at point P. Thus for a coordinate transformation that transforms the metric

$$(ds)^2 = g_{\alpha\beta}(x)dx^\alpha dx^\beta \tag{1.203}$$

into a Euclidean form,

$$(ds)^2 = (dx'^1)^2 + (dx'^2)^2 + \cdots + (dx'^N)^2 = \delta_{\mu\nu}dx'^\mu dx'^\nu, \tag{1.204}$$

the new metric functions $g'_{\alpha\beta}(x)$ must satisfy the conditions

$$g'_{\alpha\beta}(x_P^1, x_P^2, x_P^3 \ldots x_P^N) = \delta_{\mu\nu}, \tag{1.205}$$

$$\frac{\partial g'_{\alpha\beta}(x')}{\partial x'^\kappa}\bigg|_{x_P^1, x_P^2, x_P^3 \ldots x_P^N} = 0, \tag{1.206}$$

for local Cartesian coordinates so that one can write

$$g'_{\mu\nu}(x') = \delta_{\mu\nu} + O\left[\left(x' - x_p'\right)^2\right]. \tag{1.207}$$

Example 1.9. Let us reconsider the metric for two-dimensional sphere defined by

$$(ds)^2 = g_{\alpha\beta}(x^1, x^2)dx^\alpha dx^\beta, \tag{1.208}$$

where

$$g_{11} = \frac{a^2 - (x^2)^2}{a^2 - (x^1)^2 - (x^2)^2} = 1 + \frac{(x^1)^2}{a^2 - (x^1)^2 - (x^2)^2},$$

$$g_{22} = \frac{a^2 - (x^1)^2}{a^2 - (x^1)^2 - (x^2)^2} = 1 + \frac{(x^2)^2}{a^2 - (x^1)^2 - (x^2)^2}, \tag{1.209}$$

$$g_{12} = g_{21} = \frac{x^1 x^2}{a^2 - (x^1)^2 - (x^2)^2},$$

where

$$(x^3)^2 = a^2 - (x^1)^2 - (x^2)^2. \tag{1.210}$$

Note that (x^1, x^2) corresponds to the usual Cartesian coordinates (x, y). Let us pick a point P with coordinates $\left(x_p^1, x_p^2\right)$ on the surface of the sphere so that we may introduce the transformation defined by

$$x^1 = x'^1 - x_p^1,\ x^2 = x'^2 - x_p^2 \Rightarrow dx^1 = dx'^1,\ dx^2 = dx'^2, \tag{1.211}$$

and express the metric as

$$(ds)^2 = g_{11}'(dx'^1)^2 + g_{22}'(dx^2)^2 + g_{12}'dx'^1 dx'^2 + g_{21}'dx'^2 dx'^1, \tag{1.212}$$

where

$$g_{11}'\left(x'^1, x'^2\right) = 1 + \left(x'^1 - x_p^1\right)^2 \Big/ f\left(x'^1, x'^2\right),$$

$$g_{22}'\left(x'^1, x'^2\right) = 1 + \left(x'^2 - x_p^2\right)^2 \Big/ f\left(x'^1, x'^2\right), \tag{1.213}$$

$$g_{12}'\left(x'^1, x'^2\right) = g_{21}'\left(x'^1, x'^2\right) = \left(x'^1 - x_p^1\right)\left(x'^2 - x_p^2\right)^2 \Big/ f\left(x'^1, x'^2\right),$$

and

$$f(x'^1, x'^2) = a^2 - \left(x'^1 - x_p^1\right)^2 - \left(x'^2 - x_p^2\right)^2.$$

Show that the metric for a two-dimensional sphere is locally Cartesian at point P and can be written in Euclidean form:

$$(ds)^2 = \delta_{\kappa\delta}dx'^\kappa dx'^\delta. \tag{1.214}$$

Solution: For a local Cartesian, one must be able to show that at point P,

$$g'_{\alpha\beta}(x_P^1, x_P^2, x_P^3 \ldots x_P^N) = \delta_{\alpha\beta}, \left. \frac{\partial g'_{\alpha\beta}(x')}{\partial x'^\kappa} \right|_{x_P^1, x_P^2, x_P^3 \ldots x_P^N} = 0. \tag{1.215}$$

Using the metric tensor elements, we find

$$g_{11}(x_p^1, x_p^2) = g_{22}(x_p^1, x_p^2) = 1, \; g_{12}(x_p^1, x_p^2) = 0 \tag{1.216}$$

and

$$\left. \frac{\partial g_{11}(x'^1, x'^2)}{\partial x'^1} \right|_{x_p^1, x_p^2} = \frac{\partial}{\partial x'^1}\left[\left(x'^1 - x_p^1\right)^2 \Big/ f\left(x'^1, x'^2\right) \right]$$

$$= \left[2\left(x'^1 - x_p^1\right) \Big/ f\left(x'^1, x'^2\right) + 2\left(x'^1 - x_p^1\right)^3 \Big/ f^2\left(x'^1, x'^2\right) \right]_{x_p^1, x_p^2} = 0. \tag{1.217}$$

Similarly, one can easily show that

$$\left. \frac{\partial g_{11}(x'^1, x'^2)}{\partial x'^2} \right|_{x_p^1, x_p^2} = \frac{\partial}{\partial x'^2}\left[\left(x'^1 - x_p^1\right)^2 \Big/ f(x'^1, x'^2) \right]_{x_p^1, x_p^2}$$

$$= \left[2\left(x'^1 - x_p^1\right)^2\left(x'^2 - x_p^2\right) \Big/ f^2(x'^1, x'^2) \right]_{x_p^1, x_p^2} = 0, \tag{1.218}$$

$$\left. \frac{\partial g_{22}(x'^1, x'^2)}{\partial x'^1} \right|_{x_p^1, x_p^2} = \frac{\partial}{\partial x'^1}\left[\left(x'^2 - x_p^2\right)^2 \Big/ f(x'^1, x'^2) \right]_{x_p^1, x_p^2}$$

$$= \left[2\left(x'^1 - x_p^1\right)\left(x'^2 - x_p^2\right)^2 \Big/ f^2(x'^1, x'^2) \right]_{x_p^1, x_p^2} = 0, \tag{1.219}$$

$$\left. \frac{\partial g_{22}(x'^1, x'^2)}{\partial x'^2} \right|_{x_p^1, x_p^2} = \frac{\partial}{\partial x'^2}\left[\left(x'^2 - x_p^2\right)^2 \Big/ f\left(x'^1, x'^2\right) \right]_{x_p^1, x_p^2}$$

$$= \left[2\left(x'^2 - x_p^2\right) \Big/ f\left(x'^1, x'^2\right) + 2\left(x'^2 - x_p^2\right)^3 \Big/ f^2\left(x'^1, x'^2\right) \right]_{x_p^1, x_p^2} \tag{1.220}$$

$$= 0,$$

$$\left. \frac{\partial g_{12}(x'^1, x'^2)}{\partial x'^1} \right|_{x_p^1, x_p^2} = \frac{\partial}{\partial x'^1}\left[2\left(x'^1 - x_p^1\right)\left(x'^2 - x_p^2\right)^2 \Big/ f(x'^1, x'^2) \right]_{x_p^1, x_p^2} = 0,$$

$$\left. \frac{\partial g_{12}(x'^1, x'^2)}{\partial x'^2} \right|_{x_p^1, x_p^2} = \frac{\partial}{\partial x'^2}\left[2\left(x'^1 - x_p^1\right)\left(x'^2 - x_p^2\right)^2 \Big/ f(x'^1, x'^2) \right]_{x_p^1, x_p^2} = 0, \tag{1.221}$$

which leads to

$$(ds)^2 = \delta_{\kappa\delta} dx'^{\kappa} dx'^{\delta}. \tag{1.222}$$

We have seen that it is possible to find a local Cartesian coordinates at point P for an arbitrary point P in an N-dimensional Riemannian manifold where the metric takes a Euclidean form,

$$(ds)^2 = (dx'^1)^2 + (dx'^2)^2 + \cdots (dx'^N)^2.$$

The space defined by this equation is called the *tangent space* at point $P(T_p)$. We can visualize the tangent space for a two-dimensional manifold in a three-dimensional Euclidean space. This tangent space, for example, for a two-dimensional sphere forms a tangent plane at point P. This is shown in figure 1.27, where we see a two-dimensional tangent plane (shown by the red rectangular grid lines) at point P in the two-dimensional sphere. In the tangent plane shown in figure 1.27, note that

$$(ds)^2 = (dx'^1)^2 + (dx'^2)^2, \tag{1.223}$$

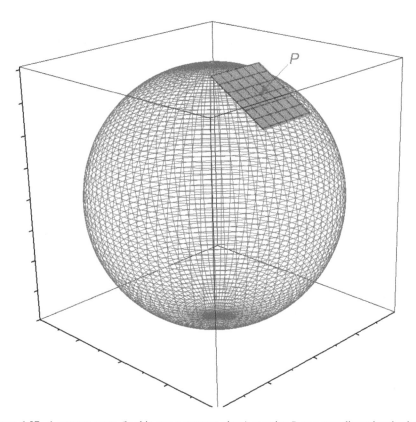

Figure 1.27. A tangent space (in this case a tangent plane) at point P on a two-dimensional sphere.

as can be seen from the rectangular shape of the grids. We can show this using the function that defines the surface of a sphere. The points on the surface of the sphere shown in figure 1.27 are defined by the function

$$g(x^1, x^2, x^3) = (x^1)^2 + (x^2)^2 + (x^3)^3 = a^2. \tag{1.224}$$

Taking the total differential for this function,

$$dg = \frac{\partial g}{\partial x^1}dx^1 + \frac{\partial g}{\partial x^2}dx^2 + \frac{\partial g}{\partial x^3}dx^3 = 0$$

$$\Rightarrow dg = \left(\frac{\partial g}{\partial x^1}\hat{x} + \frac{\partial g}{\partial x^2}\hat{y} + \frac{\partial g}{\partial x^3}\hat{z}\right) \cdot (dx^1\hat{x} + dx^2\hat{y} + dx^3\hat{z}) = 0 \tag{1.225}$$

$$dg = \vec{A}(x^1, x^2, x^3) \cdot d\vec{r}(x^1, x^2, x^3) = 0,$$

where

$$\vec{A}(x^1, x^2, x^3) = \nabla g = \frac{\partial g}{\partial x^1}\hat{x} + \frac{\partial g}{\partial x^2}\hat{y} + \frac{\partial g}{\partial x^3}\hat{z}$$

$$\frac{\partial g}{\partial x^1} = 2x^1, \frac{\partial g}{\partial x^2} = 2x^2, \frac{\partial g}{\partial x^3} = 2x^3. \tag{1.226}$$

Note equation (1.225) shows that the vector $\vec{A}(x^1, x^2, x^3)$ is normal to the surface at the point with coordinates (x^1, x^2, x^3). Let this point be P with coordinates $\vec{r}_p = \left(x_p^1, x_p^2, x_p^3\right)$, so that

$$\vec{A}\left(x_p^1, x_p^2, x_p^3\right) = \left[\frac{\partial g}{\partial x^1}\hat{x} + \frac{\partial g}{\partial x^2}\hat{y} + \frac{\partial g}{\partial x^3}\hat{z}\right]_{x_p^1, x_p^2, x_p^3} \tag{1.227}$$

$$\Rightarrow \vec{A}\left(x_p^1, x_p^2, x_p^3\right) = m_1\hat{x} + m_2\hat{y} + m_3\hat{z},$$

where

$$m_1 = 2x_p^1, m_2 = 2x_p^2, m_3 = 2x_p^3. \tag{1.228}$$

Now consider another neighboring point Q with coordinates $\vec{r} = (x^1, x^2, x^3)$ at the same plane that is tangent to the surface at point P. Then the vector on this plane is given by

$$\Delta\vec{r} = \vec{r} - \vec{r}_p = \left(x^1 - x_p^1\right)\hat{x} + \left(x^2 - x_p^2\right)\hat{y} + \left(x^3 - x_p^3\right)\hat{z} \tag{1.229}$$

and the equation of the tangent plane is determined by

$$\Delta g = \vec{A}\left(x_p^1, x_p^2, x_p^3\right) \cdot \Delta\vec{r}(x^1, x^2, x^3)$$

$$= m_1\left(x^1 - x_p^1\right) + m_2\left(x^2 - x_p^2\right) + m_3\left(x^3 - x_p^3\right) = 0. \tag{1.230}$$

$$\Rightarrow x^3 = -\frac{1}{m_3}\left[m_1\left(x^1 - x_p^1\right) + m_2\left(x^2 - x_p^2\right)\right] + x_p^3,$$

which becomes

$$x^3 = x_p^3 \Rightarrow dx^3 = dx_p^3$$

at the tangent point P. Then the metric in the tangent space becomes

$$(ds)^2 = (dx^1)^2 + (dx^2)^2 + (dx^2)^3 = (dx^1)^2 + (dx^2)^2,$$

which shows it is locally Cartesian.

1.7 The signature of a manifold

Consider an arbitrary point P in a pseudo-Riemannian manifold described by the coordinates x^α. Let the coordinates x'^α be a local Cartesian at this point. Then the metric for this transformation, given by

$$g'_{\mu\nu}(x') = g_{\alpha\beta}(x')\frac{\partial x^\alpha}{\partial x'^\mu}\frac{\partial x^\beta}{\partial x'^\nu}, \tag{1.231}$$

at point P becomes

$$g'_{\mu\nu}(x_p) = g_{\alpha\beta}(x_p)\frac{\partial x^\alpha}{\partial x'^\mu}\bigg|_{x_p} \frac{\partial x^\beta}{\partial x'^\nu}\bigg|_{x_p}. \tag{1.232}$$

Since the metric can be negative for pseudo-Riemannian manifolds, the conditions in equations (1.205) and (1.206) for a local Cartesian transformation at P must be written as

$$g'_{\mu\nu}(x_p) = \delta_{\mu\nu}\lambda_\mu, \quad \frac{\partial g'_{\mu\nu}(x')}{\partial x'^\sigma}\bigg|_{x'=x_p} = 0, \tag{1.233}$$

where $\lambda_\mu = \pm$. Therefore, for pseudo-Riemannian manifolds, the transformation metric in equation (1.232) for a local Cartesian at point P becomes

$$g'_{\mu\nu}(x_p) = \delta_{\mu\nu}\lambda_\mu = g_{\alpha\beta}(x_p)\frac{\partial x^\alpha}{\partial x'^\mu}\frac{\partial x^\beta}{\partial x'^\nu}\bigg|_{x_p}$$

$$\Rightarrow g'_{\mu\nu}(x_p) = \delta_{\mu\nu}\lambda_\mu = \frac{\partial x^\alpha}{\partial x'^\mu}\bigg|_{x_p} g_{\alpha\beta}(x_p)\frac{\partial x^\beta}{\partial x'^\nu}\bigg|_{x_p}. \tag{1.234}$$

We recall from multiplication of two N-dimensional square matrices

$$C = AB, \tag{1.235}$$

the element C_i^j (in row i and column j) in terms of A_i^k (the element in column k and row i) and B_k^j (the element in column j and row k) is given by

$$C_i^j = A_i^1 B_1^j + A_i^2 B_2^j + \cdots A_i^N B_N^j = \sum_{k=1}^{N} A_i^k B_k^j = A_i^k B_k^j. \tag{1.236}$$

In view of equation (1.236), equation (1.234) can be put in the form

$$g'_{\mu\nu}(x_p) = \delta_{\mu}\lambda_{\mu} = \Lambda_{\mu}^{\alpha}(x_p)g_{\alpha\beta}(x_p)\Lambda_{\nu}^{'\beta}(x_p),$$
(1.237)

where $\Lambda_{\mu}^{\alpha}(x_p)$, $\Lambda_{\nu}^{'\beta}(x_p)$, $g'_{\mu\nu}(x_p)$, and $g_{\alpha\beta}(x_p)$ are the elements of the matrices defined by

$$\Lambda = \left[\frac{\partial x^{\alpha}}{\partial x'^{\mu}}\right]_{x_p}, \quad \Lambda' = \left[\frac{\partial x^{\beta}}{\partial x'^{\nu}}\right]_{x_p},$$

$$G' = \left[g'_{\mu\nu}(x_p)\right] = \left[\delta_{\mu}\lambda_{\mu}\right], \quad G = \left[g_{\alpha\beta}(x_p)\right].$$
(1.238)

Using these matrices, one can rewrite equation (1.237) as

$$G' = \Lambda G \Lambda',$$
(1.239)

Since the metric $((ds)^2 = (ds')^2)$ is invariant under a coordinate transformation, we must then have an orthogonal transformation matrix at the local point P, which requires

$$\Lambda = \Lambda'^{-1} = \Lambda'^{T}.$$
(1.240)

One can then rewrite equation (1.237) as

$$G' = \Lambda'^{-1}G\Lambda' = \Lambda'^{T}G\Lambda'.$$
(1.241)

Let us see what equation (1.241) means for a two-dimensional manifold $N = 2$. We note that for a two-dimensional manifold $x^{\alpha} = x^1(x'^1, x'^2), x^2(x'^1, x'^2)$ and

$$G' = \begin{bmatrix} \lambda_1 & 0 \\ 0 & \lambda_2 \end{bmatrix}, \quad G = \begin{bmatrix} g_{11} & g_{12} \\ g_{21} & g_{22} \end{bmatrix}_{x_p^1, x_p^2},$$

$$\Lambda' = \begin{vmatrix} \dfrac{\partial x^1}{\partial x'^1} & \dfrac{\partial x^1}{\partial x'^2} \\ \dfrac{\partial x^2}{\partial x'^1} & \dfrac{\partial x^2}{\partial x'^1} \end{vmatrix}_{x_p^1, x_p^2} \Rightarrow \Lambda'^{T} = \begin{vmatrix} \dfrac{\partial x^1}{\partial x'^1} & \dfrac{\partial x^2}{\partial x'^1} \\ \dfrac{\partial x^1}{\partial x'^2} & \dfrac{\partial x^2}{\partial x'^1} \end{vmatrix}_{x_p^1, x_p^2}.$$
(1.242)

Thus equation (1.241) becomes

$$G' = \Lambda'^{T}G\Lambda' = \begin{bmatrix} \lambda_1 & 0 \\ 0 & \lambda_2 \end{bmatrix}$$
(1.243)

$$= \left\{ \begin{bmatrix} \dfrac{\partial x^1}{\partial x'^1} & \dfrac{\partial x^2}{\partial x'^1} \\ \dfrac{\partial x^1}{\partial x'^2} & \dfrac{\partial x^2}{\partial x'^1} \end{bmatrix} \begin{bmatrix} g_{11} & g_{12} \\ g_{21} & g_{22} \end{bmatrix} \begin{bmatrix} \dfrac{\partial x^1}{\partial x'^1} & \dfrac{\partial x^1}{\partial x'^2} \\ \dfrac{\partial x^2}{\partial x'^1} & \dfrac{\partial x^2}{\partial x'^1} \end{bmatrix} \right\}_{x_p^1, x_p^2}.$$
(1.244)

We recall the similarity transformation for the matrix M, which is given by

$$D = C^{-1}MC, \tag{1.245}$$

where C is a matrix whose columns are the eigenvectors of the eigenvalue equation for matrix M, and the matrix D is a diagonal matrix whose diagonal elements are the eigenvalues of M. Also we know that the transformation is orthogonal:

$$C^{-1} = C^{\mathrm{T}} \Rightarrow D \equiv C^{\mathrm{T}}MC. \tag{1.246}$$

Therefore, since the matrix G' is a diagonal matrix, the transformation from the x^α coordinates to the local Cartesian coordinates x'^α, in equation (1.241), is a *similarity transformation*. This means matrix Λ' forms the eigenvector matrix and G' forms the eigenvalue matrix. Now let us consider the inverse transformation, which means from x'^α to x^α. For this inverse transformation, one can then write for the metric tensor

$$G = \Lambda^{-1}G'\Lambda = \Lambda^{\mathrm{T}}G'\Lambda, \tag{1.247}$$

where

$$\Lambda = \left[\frac{\partial x'^\beta}{\partial x^\nu}\right]_{x_p} \Rightarrow \Lambda^{-1} = \Lambda^{\mathrm{T}} = \left[\frac{\partial x'^\nu}{\partial x^\beta}\right]_{x_p}. \tag{1.248}$$

Upon substituting equation (1.241), for G', into equation (1.247), we find

$$G = \Lambda^{\mathrm{T}}\Lambda'^T G\Lambda'\Lambda, \tag{1.249}$$

which clearly indicates that, for the local Cartesian coordinate transformation, we must then have

$$\Lambda'\Lambda = \Lambda^{\mathrm{T}}\Lambda'^T = I \Rightarrow \Lambda' = \Lambda^{-1}, \Lambda^{\mathrm{T}} = (\Lambda'^T)^{-1}. \tag{1.250}$$

Since, for a similarity transformation, $\Lambda^{-1} = \Lambda^{\mathrm{T}}$ and $(\Lambda')^{-1} = (\Lambda')^{\mathrm{T}}$, one can easily find

$$\Lambda' = \Lambda^{\mathrm{T}} \Rightarrow \left[\frac{\partial x^\beta}{\partial x'^\nu}\bigg|_{x_p}\right] = \left[\frac{\partial x'^\beta}{\partial x^\nu}\bigg|_{x_p}\right]^{\mathrm{T}}$$

$$\Rightarrow \left[\frac{\partial x^\beta}{\partial x'^\nu}\bigg|_{x_p}\right] = \left[\frac{\partial x'^\nu}{\partial x^\beta}\bigg|_{x_p}\right]. \tag{1.251}$$

This can only be true if the transformation from $x^\alpha \to x'^\alpha$ is linear, which means all the elements of the transformation matrix Λ^α_β at point P must be a constant, and we can write

$$x'^\alpha = \Lambda^\alpha_\beta(x_p)x^\beta. \tag{1.252}$$

Now let us consider this condition for the two-dimensional manifold in equation (1.244). Noting that

$$x'^1 = \Lambda_1^1(x_p)x^1 + \Lambda_2^1(x_p)x^2, \ x'^2 = \Lambda_1^2(x_p)x^1 + \Lambda_2^2(x_p)x^2$$

$$\Rightarrow \Lambda = \left[\frac{\partial x'^\beta}{\partial x^\nu}\bigg|_{x_p}\right] = \begin{bmatrix} \dfrac{\partial x'^1}{\partial x^1} & \dfrac{\partial x'^1}{\partial x^2} \\[2mm] \dfrac{\partial x'^2}{\partial x^1} & \dfrac{\partial x'^2}{\partial x^2} \end{bmatrix}_{x_p^1, x_p^2} = \begin{bmatrix} \Lambda_1^1 & \Lambda_2^1 \\ \Lambda_1^2 & \Lambda_2^2 \end{bmatrix}_{x_p^1, x_p^2} \quad (1.253)$$

$$\Rightarrow \Lambda^T = \begin{bmatrix} \Lambda_1^1 & \Lambda_1^2 \\ \Lambda_2^1 & \Lambda_2^2 \end{bmatrix},$$

one can rewrite equation (1.244) as

$$G' = \Lambda^T G \Lambda \Rightarrow \begin{bmatrix} \lambda_1 & 0 \\ 0 & \lambda_2 \end{bmatrix}$$

$$= \left\{ \begin{bmatrix} \Lambda_1^1 & \Lambda_1^2 \\ \Lambda_2^1 & \Lambda_2^2 \end{bmatrix} \times \begin{bmatrix} g_{11} & g_{12} \\ g_{21} & g_{22} \end{bmatrix} \begin{bmatrix} \Lambda_1^1 & \Lambda_2^1 \\ \Lambda_1^2 & \Lambda_2^2 \end{bmatrix} \right\}_{x_p^1, x_p^2}. \quad (1.254)$$

Recalling that the matrix Λ is a matrix constructed from the eigenvectors and G' is the eigenvalues for the matrix G eigenvalue equation, one can divide the eigenvectors by the square root of the absolute value of the corresponding eigenvalues and write

$$G' = \begin{bmatrix} \pm 1 & 0 \\ 0 & \pm 1 \end{bmatrix}$$

$$= \left\{ \begin{bmatrix} \dfrac{\Lambda_1^1}{\sqrt{|\lambda_1|}} & \dfrac{\Lambda_1^2}{\sqrt{|\lambda_1|}} \\[3mm] \dfrac{\Lambda_2^1}{\sqrt{|\lambda_2|}} & \dfrac{\Lambda_2^2}{\sqrt{|\lambda_2|}} \end{bmatrix} \times \begin{bmatrix} g_{11} & g_{12} \\ g_{21} & g_{22} \end{bmatrix} \begin{bmatrix} \dfrac{\Lambda_1^1}{\sqrt{|\lambda_1|}} & \dfrac{\Lambda_2^1}{\sqrt{|\lambda_2|}} \\[3mm] \dfrac{\Lambda_1^2}{\sqrt{|\lambda_1|}} & \dfrac{\Lambda_2^2}{\sqrt{|\lambda_2|}} \end{bmatrix} \right\}_{x_p^1, x_p^2}, \quad (1.255)$$

where we take into account the fact that the eigenvalues for λ_μ could be plus or minus for a pseudo-Riemannian manifold. This can be generalized for an N-dimensional pseudo-Riemannian manifold:

$$G' = \begin{bmatrix} \lambda_1 & 0 & \cdots & 0 \\ 0 & \lambda_2 & \cdots & 0 \\ \vdots & \cdot & \cdot & \vdots \\ 0 & 0 & \cdots & \lambda_N \end{bmatrix}. \quad (1.256)$$

If we scale the coordinates x'^α by these eigenvalues (i.e., $x'^\alpha \to x'^\alpha/\sqrt{|\lambda_\alpha|}$), for the metric tensor, G', we can write

$$G' = \begin{bmatrix} \pm 1 & 0 & \cdots & 0 \\ 0 & \pm 1 & \cdots & 0 \\ \vdots & & \cdot & \vdots \\ 0 & 0 & \cdots & \pm 1 \end{bmatrix} = [\eta_{\alpha\beta}]. \tag{1.257}$$

Thus at any arbitrary point P in a pseudo-Riemannian manifold it is always possible to find a coordinate system $x^{'\alpha}$ such that in the neighborhood of P, we have

$$g'_{\alpha\beta}(x') = \eta_{\alpha\beta} + O\left[\left(x' - x'_p\right)^2\right]. \tag{1.258}$$

Suppose the number of positive entries in the matrix $[\eta_{\alpha\beta}]$ be N_+ and that of the negative entries be N_-. The *signature* of the manifold, which we represent by ξ, is defined as

$$\xi = N_+ - N_-. \tag{1.259}$$

For example, for the Minkowski space-time manifold where the metric is given by

$$(ds)^2 = (dx^{'1})^2 - (dx^{'2})^2 - (dx^{'3})^2 - (dx^{'4})^2, \tag{1.260}$$

with $(x^{'1}, x^{'2}, x^{'3}, x^{'4}) \to (ct, x, y, z)$, we have

$$[\eta_{\alpha\beta}] = \begin{bmatrix} 1 & 0 & 0 & 0 \\ 0 & -1 & 0 & 0 \\ 0 & 0 & -1 & 0 \\ 0 & 0 & 0 & -1 \end{bmatrix}, \tag{1.261}$$

$N_+ = 1$, $N_- = 3$, and one finds

$$\xi = N_+ - N_- = -2, \tag{1.262}$$

which is the signature for the Minkowski space-time manifold.

1.8 *N*-dimensional volume without a constraint

In section 1.5 we saw that N-volume in an N-dimensional manifold is given by

$$d^N V = \sqrt{|g_{11}g_{22}\cdots g_{NN}|} \, dx^1 dx^2 \cdots dx^N. \tag{1.263}$$

We found this volume when the coordinates are orthogonal. Next, we will show how this same expression is valid even for non-orthogonal coordinates. The key is to find a local Cartesian coordinates transformation. Before we go into more detail, it is important to show how the volume in one coordinate system is related to another coordinate by the Jacobian of the coordinate transformation. To this end, we shall consider the following example.

Example 1.10. We recall the infinitesimal volume of a sphere in Cartesian $(x^{'1}, x^{'2}, x^{'3}) \to (x, y, z)$ and in spherical $(x^1, x^2, x^3) \to (r, \theta, \varphi)$ coordinates (see figure 1.28) is related by

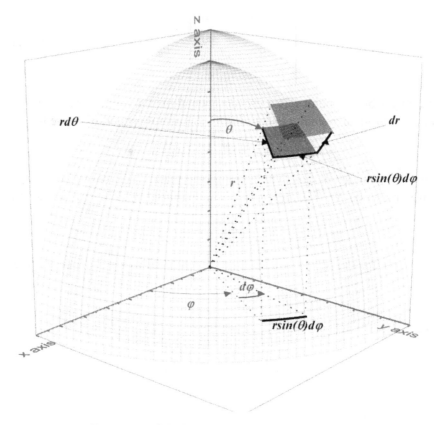

Figure 1.28. Infinitesimal volume in spherical coordinates.

$$d^3V = dxdydz = r \sin(\theta)drd\theta d\varphi,$$
$$\Rightarrow d^3V = dx'^1 dx'^2 dx'^3 = (x^1)^2 \sin(x^2) dx^1 dx^2 dx^3, \tag{1.264}$$

where

$$x = r \sin(\theta)\cos(\varphi), \ y = r \sin(\theta)\sin(\varphi), \ z = \cos(\theta),$$
$$\Rightarrow x'^1 = x^1 \sin(x^2)\cos(x^3), \ x'^2 = x^1 \sin(x^2)\sin(x^3), \tag{1.265}$$
$$x'^3 = x^1 \cos(x^2).$$

For this coordinate transformation, show that the Jacobian is given by

$$J = \det\left[\frac{\partial x'^\alpha}{\partial x^\beta}\right] = \begin{vmatrix} \dfrac{\partial x'^1}{\partial x^1} & \dfrac{\partial x'^1}{\partial x^2} & \dfrac{\partial x'^1}{\partial x^3} \\[2mm] \dfrac{\partial x'^2}{\partial x^1} & \dfrac{\partial x'^2}{\partial x^2} & \dfrac{\partial x'^2}{\partial x^3} \\[2mm] \dfrac{\partial x'^3}{\partial x^1} & \dfrac{\partial x'^3}{\partial x^2} & \dfrac{\partial x'^3}{\partial x^3} \end{vmatrix} = (x^1)^2 \sin(x^2), \tag{1.266}$$

and the three-volume by

$$d^3V = dx'^1 dx'^2 dx'^3 = |J| \, dx^1 dx^2 dx^3. \qquad (1.267)$$

Solution: Noting that

$$\frac{\partial x'^1}{\partial x^1} = \frac{\partial}{\partial x^1}[x^1 \sin(x^2) \cos(x^3)] = \sin(x^2) \cos(x^3),$$

$$\frac{\partial x'^1}{\partial x^2} = \frac{\partial}{\partial x^2}[x^1 \sin(x^2) \cos(x^3)] = x^1 \cos(x^2) \cos(x^3), \qquad (1.268)$$

$$\frac{\partial x'^1}{\partial x^3} = \frac{\partial}{\partial x^3}[x^1 \sin(x^2) \cos(x^3)] = -x^1 \sin(x^2) \sin(x^3),$$

$$\frac{\partial x'^2}{\partial x^1} = \frac{\partial}{\partial x^1}[x^1 \sin(x^2) \sin(x^3)] = \sin(x^2) \sin(x^3),$$

$$\frac{\partial x'^2}{\partial x^2} = \frac{\partial}{\partial x^2}[x^1 \sin(x^2) \sin(x^3)] = x^1 \cos(x^2) \sin(x^3), \qquad (1.269)$$

$$\frac{\partial x'^2}{\partial x^3} = \frac{\partial}{\partial x^3}[x^1 \sin(x^2) \sin(x^3)] = x^1 \sin(x^2) \cos(x^3),$$

$$\frac{\partial x'^3}{\partial x^1} = \frac{\partial}{\partial x^1}[x^1 \cos(x^2)] = \cos(x^2),$$

$$\frac{\partial x'^3}{\partial x^2} = \frac{\partial}{\partial x^2}[x^1 \cos(x^2)] = -x^1 \sin(x^2), \qquad (1.270)$$

$$\frac{\partial x'^3}{\partial x^3} = \frac{\partial}{\partial x^3}[x^1 \cos(x^2)] = 0,$$

one can write the Jacobian as

$$J = \begin{vmatrix} \sin(x^2)\cos(x^3) & x^1\cos(x^2)\cos(x^3) & -x^1\sin(x^2)\sin(x^3) \\ \sin(x^2)\sin(x^3) & x^1\cos(x^2)\sin(x^3) & x^1\sin(x^2)\cos(x^3) \\ \cos(x^2) & -x^1\sin(x^2) & 0 \end{vmatrix}$$

$$= \begin{vmatrix} \cos(x^2) & -x^1\sin(x^2) & 0 \\ \sin(x^2)\cos(x^3) & x^1\cos(x^2)\cos(x^3) & -x^1\sin(x^2)\sin(x^3) \\ \sin(x^2)\sin(x^3) & x^1\cos(x^2)\sin(x^3) & x^1\sin(x^2)\cos(x^3) \end{vmatrix} \qquad (1.271)$$

$$\Rightarrow J = \left\{ (x^1)^2 \sin(x^2) \cos^2(x^2)[\cos^2(x^3) + \sin^2(x^3)] \right.$$
$$\left. + (x^1)^2 \sin^3(x^2)[\cos^2(x^3) + \sin^2(x^3)] \right\}$$

$$\Rightarrow J = (x^1)^2 \sin(x^2).$$

Note that the minus sign in the determinant is due to an interchange of the first and the third rows. Therefore, the volume in the two coordinates,

$$d^3V = dxdydz = r\sin(\theta)drd\theta d\varphi,$$
$$d^3V = dx'^1 dx'^2 dx'^3 = (x^1)^2 \sin(x^2)dx^1 dx^2 dx^3, \tag{1.272}$$

can be written as

$$d^3V = dx'^1 dx'^2 dx'^3 = |J| \, dx^1 dx^2 dx^3. \tag{1.273}$$

Note that in example 1.9 the coordinates $(x'^1, x'^2, x'^3) \to (x, y, z)$ form a local Cartesian for the sphere. Therefore, based on the result in example 1.9 for an N-dimensional (pseudo-)Riemannian manifold with non-orthogonal coordinates, one can find local Cartesian coordinates such that

$$d^N V = dx'^1 dx'^2 \cdots dx'^N = |J| \, dx^1 dx^2 \cdots dx^N. \tag{1.274}$$

In the last section, for local Cartesian coordinates in an N-dimensional (pseudo-)Riemannian manifold, we showed that

$$G' = \left[g'_{\alpha\beta} \right] = \begin{bmatrix} g'_{11} & 0 & \cdots & 0 \\ 0 & g'_{22} & \cdots & 0 \\ \vdots & \cdots & \cdots & \vdots \\ 0 & 0 & \cdots & g'_{NN} \end{bmatrix} = [\eta_{\alpha\beta}] = \begin{bmatrix} 1 & 0 & 0 & 0 \\ 0 & -1 & 0 & 0 \\ 0 & 0 & -1 & 0 \\ 0 & 0 & 0 & -1 \end{bmatrix}, \tag{1.275}$$

where $x'^\alpha \to x'^\alpha / \sqrt{|\lambda_\alpha|}$ for a (pseudo-)Riemannian manifold. Using the notation from the previous section,

$$\Lambda' = \left[\frac{\partial x^\beta}{\partial x'^\alpha} \right]_{x_p} \Rightarrow \Lambda'^T = \left[\frac{\partial x^\alpha}{\partial x'^\beta} \right]_{x_p}, \tag{1.276}$$

and recalling that

$$\Lambda^T = \Lambda^{-1} = \Lambda' = \left[\frac{\partial x^\alpha}{\partial x'^\beta} \bigg|_{x_p} \right] \Rightarrow \Lambda = \Lambda'^T = \left[\frac{\partial x^\beta}{\partial x'^\alpha} \bigg|_{x_p} \right], \tag{1.277}$$

one can write for the Jacobian

$$J = \det |\Lambda| = \det \left[\frac{\partial x'^\alpha}{\partial x^\beta} \bigg|_{x_p} \right] = \det \left[\frac{\partial x^\beta}{\partial x'^\alpha} \bigg|_{x_p} \right]$$

$$J = \det |\Lambda| = \det \left[\frac{\partial x^\alpha}{\partial x'^\beta} \bigg|_{x_p} \right] = \det [(\Lambda')^{-1}]. \tag{1.278}$$

Furthermore, we know

$$\Lambda\Lambda^{-1} = \Lambda^{-1}\Lambda = I \Rightarrow \det |\Lambda^{-1}| = \frac{1}{\det |\Lambda|},$$

so that using the result in equation (1.278), one finds

$$\det | \Lambda^{-1} | = \frac{1}{J}. \tag{1.279}$$

We recall for a local Cartesian coordinate transformation

$$G' = (\Lambda')^{-1} G \Lambda' = (\Lambda^{\mathrm{T}})^{-1} G \Lambda' = (\Lambda^{-1})^{-1} G \Lambda' = \Lambda G \Lambda'$$
$$\Rightarrow \det | G' | = \det | \Lambda G \Lambda' | = \det | \Lambda | \det | G | \det | \Lambda' | \tag{1.280}$$

$$= \det | \Lambda | \det | \Lambda' | \det (G) = [\det (\Lambda)]^2 \det (G)$$
$$\Rightarrow \det (G') = [\det (\Lambda)]^2 \det (G). \tag{1.281}$$

Let the determinant of G and G' be represented by g and g', respectively. Then using equation (1.279) one can write equation (1.281) as

$$g' = \frac{1}{J^2} g. \tag{1.282}$$

Since x'^{α} coordinates are locally Cartesian and the metric tensor

$$G' = \left[g'_{\alpha\beta} \right] = [\eta_{\alpha\beta}] = \begin{bmatrix} 1 & 0 & 0 & 0 \\ 0 & -1 & 0 & 0 \\ 0 & 0 & -1 & 0 \\ 0 & 0 & 0 & -1 \end{bmatrix}, \tag{1.283}$$

for the determinant of G' we always find

$$g' = \det (G') = \pm 1, \tag{1.284}$$

and equation (1.282) becomes

$$g' = \frac{1}{J^2} g \Rightarrow \pm 1 = \frac{1}{J^2} g \Rightarrow J^2 = \pm g \Rightarrow | J | = | g |, \tag{1.285}$$

and the full N-dimensional volume element $d^N V$,

$$d^N V = dx'^1 dx'^2 \cdots dx'^N = | J | dx^1 dx^2 \cdots dx^N, \tag{1.286}$$

becomes

$$d^N V = dx''^1 dx'^2 \cdots dx'^N = \sqrt{| g |} \, dx^1 dx^2 dx^3 \cdots dx^N. \tag{1.287}$$

This relation is valid for a (pseudo-)Riemannian manifold with orthogonal or non-orthogonal coordinates.

1.9 Homework assignment

Problem 1. For the coordinate transformation we considered in example 1.2,

$$x'^1 = x^1 \cos(\theta) + x^2 \sin(\theta),$$
$$x'^2 = -x^1 \sin(\theta) + x^2 \cos(\theta),$$
$$x'^3 = x^3.$$

(a) Solve for x^1, x^2 and x^3.
(b) Find the inverse transformation matrix
$$\frac{\partial x^\alpha}{\partial x'^\beta}.$$

(c) And the Jacobian, J',
$$J' = \det \left| \frac{\partial x^\alpha}{\partial x'^\beta} \right|.$$

(d) Is there any relationship between the Jacobian J in example 1.2 and J'?

Problem 2. For the coordinate transformation defined by
$$x^1 = x'^1 \sin(x'^2) \cos(x'^3),$$
$$x^2 = x'^1 \sin(x'^2) \sin(x'^3),$$
$$x^3 = x'^1 \cos(x'^2).$$

(a) Find the transformation matrix
$$\frac{\partial x^\alpha}{\partial x'^\beta}.$$

(b) Find the Jacobian
$$J' = \det \left| \frac{\partial x^\alpha}{\partial x'^\beta} \right|.$$

(c) Solve for x'^1, x'^2, and x'^3, and find the inverse transformation matrix
$$\frac{\partial x'^\alpha}{\partial x^\beta}.$$

This might be a bit challenging. You can use Mathematica.
(d) Then the Jacobian, J,

$$J = \det \left| \frac{\partial x^{'\alpha}}{\partial x^{\beta}} \right|.$$

(e) Is there any relationship between J and J'?

Note that $(x^{'1}, x^{'2}, x^{'3}) \to (r, \theta, \varphi)$ and $(x^1, x^2, x^3) \to (x, y, z)$ (the usual spherical and cylindrical coordinates).

Problem 3. Consider an observer (i.e., a man) on Earth (i.e., in the S frame), and another observer in the S' frame (Alien 2) that is moving with a speed v along the positive x-axis as shown in figure 1.29. Suddenly, an event occurs: the appearance of Alien 1. This event is recorded by the two observers on S and S' (i.e., the man and Alien 2). The man recorded this event at

$$(t, x, y, z) \to (ct, x, y, z) \to (x^0, x^1, x^2, x^3),$$

and Alien 2 recorded this event at

$$(t', x', y', z') \to (ct', x', y', z') \to (x^{'0}, x^{'1}, x^{'2}, x^{'3}).$$

These two coordinates are related by a Lorentz transformation for $x^{\alpha} \to x^{'\alpha}$,

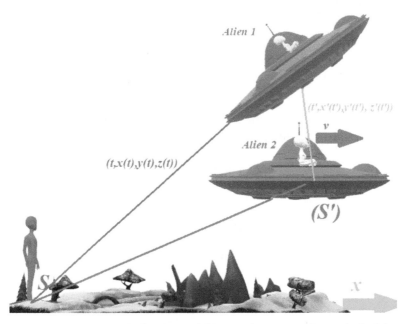

Figure 1.29. An event (appearance of Alien 1) recorded by two observers; an observer in the S frame (a man on Earth) and an observer in the S' frame (Alien 2, moving with a velocity v along the positive x axis as measured by an observer in the S frame).

$$x^0 = \gamma x'^0 + \gamma\beta x'^1, \ x^1 = \gamma\beta x'^0 + \gamma x'^1, \ x^2 = x'^2, \ x^3 = x'^3,$$

and for $x'^\alpha \rightarrow x^\alpha$,

$$x'^0 = \gamma x^0 - \gamma\beta x^1, \ x'^1 = -\gamma\beta x^0 + \gamma x^1, \ x'^2 = x^2, \ x'^3 = x^3,$$

where

$$\beta = \frac{v}{c}, \ \gamma = \frac{1}{\sqrt{1 - \dfrac{v^2}{c^2}}}.$$

Find the transformation matrices

$$[\Lambda^\kappa_{\ \alpha}] = \left[\frac{\partial x^\kappa}{\partial x'^\alpha} \right], \ [\Lambda'^\beta_{\ d}] = \left[\frac{\partial x'^\beta}{\partial x^d} \right].$$

Problem 4. For a two-dimensional sphere where a point on this sphere is described by the coordinate system (x^1, x^2) and (x'^1, x'^2), related by

$$x^1 = x'^1 \cos(x'^2), \ x^2 = x'^1 \sin(x'^2),$$

we found the metric tensor elements to be

$$g_{11}(x^1, x^2) = [a^2 - (x^2)^2]/f(x^1, x^2) = 1 + (x^1)^2/f(x^1, x^2),$$
$$g_{22}(x^1, x^2) = [a^2 - (x^1)^2]/f(x^1, x^2) = 1 + (x^2)^2/f(x^1, x^2),$$
$$g_{12}(x^1, x^2) = g_{21}(x^1, x^2) = (x^1 x^2)/f(x^1, x^2),$$

where

$$f(x^1, x^2) = a^2 - (x^1)^2 - (x^2)^2.$$

Using the relation that transforms the metric elements from the unprimed coordinate to the primed coordinates,

$$g'_{\kappa\delta}(x') = g_{\alpha\beta}(x) \frac{\partial x^\alpha}{\partial x'^\kappa} \frac{\partial x^\beta}{\partial x'^\delta},$$

show that

$$g'_{11}(x'^1, x'^2) = a^2/[a^2 - (x'^1)^2], \ g'_{22}(x'^1, x'^2) = (x'^1)^2,$$
$$g'_{12}(x'^1, x'^2) = g'_{21}(x'^1, x'^2) = 0.$$

Problem 5. For a three-dimensional sphere for the coordinate system (x^1, x^2, x^3) and (x'^1, x'^2, x'^3), related by

$$x^1 = x'^1 \sin(x'^2) \cos(x'^3), \ x^2 = \sin(x'^2) \sin(x'^3), \ x^3 = \cos(x'^2),$$

we derived the metric elements in

$$g_{11}(x^1, x^2, x^3) = [a^2 - (x^2)^2 - (x^3)^2]/f(x^1, x^2, x^3),$$
$$g_{22}(x^1, x^2, x^3) = [a^2 - (x^1)^2 - (x^3)^2]/f(x^1, x^2, x^3),$$
$$g_{33}(x^1, x^2, x^3) = [a^2 - (x^1)^2 - (x^2)^2]/f(x^1, x^2, x^3),$$
$$g_{12}(x^1, x^2, x^3) = g_{21}(x^1, x^2, x^3) = x^1 x^2/f(x^1, x^2, x^3),$$
$$g_{13}(x^1, x^2, x^3) = g_{31}(x^1, x^2, x^3) = x^1 x^3/f(x^1, x^2, x^3),$$
$$g_{23}(x^1, x^2, x^3) = g_{32}(x^1, x^2, x^3) = x^2 x^3/f(x^1, x^2, x^3),$$

where

$$f(x^1, x^2, x^3) = a^2 - (x^1)^2 - (x^2)^2 - (x^3)^2$$

and

$$g'_{11}(x'^1, x'^2, x'^3) = \frac{a^2}{a^2 - x'^{12}}, \; g'_{22}(x'^1, x'^2, x'^3) = (x'^1)^2,$$
$$g'_{33}(x'^1, x'^2, x'^3) = (x'^1)^2 \sin^2(x'^2),$$
$$g'_{12}(x'^1, x'^2, x'^3) = g'_{21}(x'^1, x'^2, x'^3) = g'_{13}(x'^1, x'^2, x'^3) = 0,$$
$$g'_{31}(x'^1, x'^2, x'^3) = g'_{23}(x'^1, x'^2, x'^3) = g'_{32}(x'^1, x'^2, x'^3) = 0.$$

Using the relation that transforms the metric elements from the unprimed coordinate to the primed coordinates,

$$g'_{\kappa d}(x') = g_{\alpha\beta}(x)\frac{\partial x^\alpha}{\partial x'^\kappa}\frac{\partial x^\beta}{\partial x'^d},$$

show that

$$g'_{12}(x'^1, x'^2, x'^3) = g'_{13}(x'^1, x'^2, x'^3) = 0$$

and

$$g'_{11}(x'^1, x'^2, x'^3) = \frac{a^2}{a^2 - (x'^1)^2}, \; g'_{22}(x'^1, x'^2, x'^3) = (x'^1)^2.$$

Problem 6. In example 1.7, express the circumference (part b) and the area (part c) in terms of the distance D (part a) and determine D for the maximum circumference and area. Find the maximum circumference and area of this sphere. Is your answer consistent with what you know about the maximum circumference and surface area of a sphere with radius a?

Problem 7.
   (a) Express the circumference, the area, and the volume in terms of D in example 1.7 and show that all have maximum values at

$$D = \frac{\pi a}{2}.$$

(b) Show that the total volume of this space is finite and is equal to

$$V = 2\pi^2 a^3.$$

Problem 8. Determine the metric for a three-dimensional sphere defined by the coordinates $(x'^1, x'^2, x'^3) \to (r, \theta, \varphi)$ with imaginary radius $a = ib$ embedded in a four-dimensional Euclidean space (see example 1.8). Then consider a two-dimensional sphere defined by $r = R$.

(a) Show that circumference C and the area A are given by $C = 2\pi R$ and $A = 4\pi R^2$, respectively.

(b) Show that the distance D from the center of the sphere to the surface is

$$D = b \sinh^{-1}(R/b).$$

(c) Show that A and V of the sphere are monotonically increasing functions.

Problem 9. For the local Cartesian coordinate transformation matrix

$$X = \left[\frac{\partial x^\beta}{\partial x'^\alpha} \Big|_{x_p} \right],$$

and inverse transformation matrix

$$X' = \left[\frac{\partial x'^\beta}{\partial x^\alpha} \Big|_{x_p} \right],$$

when

$$X' = X^{-1} = X^{\mathrm{T}},$$

we have made the statement that this can only be true if the transformation $x^\alpha \to x'^\alpha$ is linear:

$$x'^\alpha = X^\alpha_\beta x^\beta,$$

which means all the elements of the transformation matrix X (i.e., $X^\alpha_\beta = $ constant) at point P. Using a two-dimensional manifold, prove this statement.

Problem 10. By identifying a suitable coordinate transformation, show that the metric

$$(ds)^2 = (c^2 - a^2 t^2)^2 (dt)^2 - 2at\,dt\,dx - (dx)^2 - (dy)^2 - (dz)^2,$$

where a is a constant, can be reduced to a Minkowski metric:

$$(ds)^2 = (dx'^1)^2 - (dx'^2)^2 - (dx'^3)^2 - (dx'^4)^2.$$

Reference

[1] Erenso D and Montemayor V 2022 *Studies in Theoretical Physics, Volume 1: Fundamental Mathematical Models* (Bristol: IOP Publishing)

IOP Publishing

Studies in Theoretical Physics, Volume 2
Advanced mathematical methods
Daniel Erenso

Chapter 2

Vector calculus on manifolds

This chapter introduces vector calculus on manifolds. It begins with a foundational understanding of scalars and vectors on a manifold, and the discussion progresses to explain the relation of the tangent vector and basis vectors that form the building blocks for vector calculus on a manifold. The metric function and coordinate basis vectors are examined, providing insight into the geometric relationships underlying manifold calculus. The chapter explores the interplay between basis vectors and coordinate transformations, revealing the components of vectors within these transformations. A detailed investigation into the inner product of vectors and the metric tensor sheds light on the geometric structure of the manifold. We then introduce the affine connections and discuss how they transform under coordinate transformations and their relationship with the metric tensor. The chapter also explores the concept of geodesics and the intrinsic derivative of a vector along a curve, the parallel transport of vectors on a manifold, and the Euler-Lagrange equation. The chapter ends with a set of problems to solve.

2.1 Scalars and vectors on a manifold

The electric potential, which is a scalar function, for a positive point charge located at the origin forms a spherical equipotential surface. In fact, this surface is defined by

$$\phi(\vec{r}) = \frac{kQ}{r} = \frac{kQ}{\sqrt{x^2 + y^2 + z^2}} = V_0, \tag{2.1}$$

where V_0 is the electric potential on the surface of the sphere. This potential on the surface of the sphere with radius $r = a$ does not change whether we express it in Cartesian coordinates, $(x, y, z) \rightarrow (x'^1, x'^2, x'^3)$, or in spherical coordinates, $(r, \theta, \varphi) \rightarrow (x^1, x^2, x^3)$, we get the same value: $\phi(\vec{r}) = V_0$. In this particular case the potential is constant anywhere on the surface of the sphere. Instead of a point charge, if we consider a tiny spherical shell with surface charge density $\sigma(\theta, \varphi) = \sigma_0 \sin(\theta) \cos(\varphi)$ at the origin, the electric potential on the surface of the sphere defined by

doi:10.1088/978-0-7503-4861-4ch2

$$(x'^1)^2 + (x'^2)^2 + (x'^3)^2 = a^2$$
$$\Rightarrow x'^3 = \pm\sqrt{a^2 - (x'^1)^2 - (x'^2)^2}$$

(2.2)

does not remain a constant and would be a function of the coordinates (x'^1, x'^2) or (x^1, x^2). But still the value of the electric potential at a point P on the surface of the sphere is the same independent of the coordinates we chose:

$$\phi(\vec{r}) = \phi(x'^1, x'^2) = \phi(x^1, x^2).$$

The explicit form of a scalar function on a manifold depends on the set of coordinates one chooses. The value of the function at a certain point on the manifold, however, remains the same. In other words, scalar quantities are invariant under coordinate transformations:

$$\phi(x'^\alpha) = \phi(x^\alpha).$$

A vector field defined on a manifold M (for example for a two-dimensional manifold; see figure 2.1) assigns a single vector to each point P or Q in the region of the manifold. At each point on the manifold local vectors $(\vec{v}_P$ and $\vec{v}_Q)$ lie in the corresponding tangent space (T_P and T_Q) as shown for a two-dimensional manifold

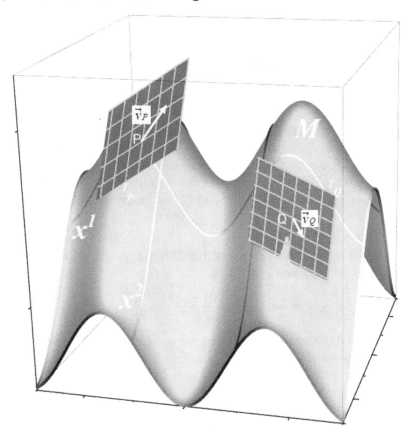

Figure 2.1. Two vectors on a manifold lie in two different tangent spaces.

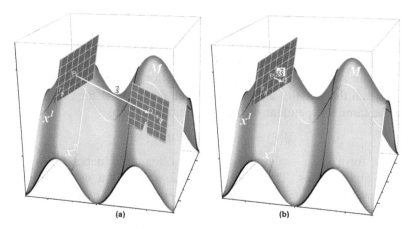

Figure 2.2. (a) A displacement vector $\vec{S}$ between two points on a two-dimensional manifold that does not lie on the manifold. (b) A displacement vector $\delta\vec{S}$ defined in the tangent space in the neighborhood of point P.

embedded in three-dimensional Euclidean space in figure 2.1. Local vectors on a manifold obey the usual vector algebra.

When we talk about position and displacement vectors on a manifold, we need to be careful. The displacement vector between two points must lie on the manifold. For example, in figure 2.2(a), the displacement vector $\vec{S}$ does not lie on the manifold. The manifold is a two-dimensional curved surface embedded in a three-dimensional Euclidean space shown by the shaded (blue) surface and described by the curved coordinates x^1 and x^2.

When we talk about position and displacement vectors on manifold, we need to be careful. *The displacement vector between two points must lie on the manifold.* For example, in figure 2.2(a), the displacement vector $\vec{S}$ does not lie on the manifold. The manifold is a two-dimensional curved surface embedded in a three-dimensional Euclidean space shown by the shaded (blue) surface and described by the curved coordinates x^1 and x^2. However, we can define an infinitesimal displacement vector, $\delta\vec{S}$, between two neighboring points P and Q that lies in the tangent space T_P.

2.2 The tangent vector

The tangent vector, $\vec{t}$, at point P on a manifold is the vector that lies in the tangent space T_p at that point P, and is given by

$$\vec{t} = \lim_{\delta u \to 0} \frac{\vec{s}(u + \delta u) - \vec{s}(u)}{\delta u} = \frac{d\vec{s}}{du}, \tag{2.3}$$

where $\delta\vec{s}$ is the infinitesimal separation vector between the point P with coordinate $(u, s(u))$ and some nearby point Q with coordinate $(u + \delta u, s(u + \delta u))$ on the curve C that lies on the manifold (see figures 2.3 and 2.4).

2.3 The basis vectors

At each point P on a manifold we can define a set of linearly independent basis vectors $\hat{e}_\alpha(x)$ for the tangent space T_P. The number of the basis vectors is equal to the

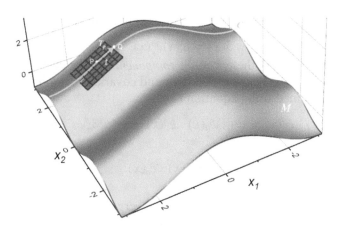

Figure 2.3. A tangent vector at point P in the tangent space.

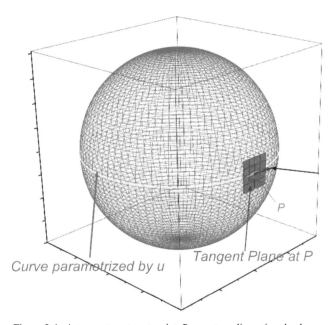

Figure 2.4. A tangent vector at point P on a two-dimensional sphere.

dimension of T_P. Any vector field $\vec{v}(x)$ at point P can then be expressed as a linear combination of these basis vectors:

$$\vec{v}(x) = v^\alpha(x)\hat{e}_\alpha(x), \tag{2.4}$$

where $v^\alpha(x)$ are known as the *contravariant components* of the vector field $\vec{v}(x)$ in the basis $\hat{e}_\alpha$. For any set of basis vectors there are a reciprocal set of basis vectors known as *the dual basis vectors* $\hat{e}^\alpha(x)$, defined by

$$\hat{e}^\alpha(x) \cdot \hat{e}_\beta(x) = \delta^\alpha_\beta. \tag{2.5}$$

The local vector field $\vec{v}(x)$ can also be expressed in terms of *the dual basis vectors* as

$$\vec{v}(x) = v_a(x)\hat{e}^a(x), \qquad (2.6)$$

where $v_a(x)$ are known as the *covariant components* of the vector field $\vec{v}(x)$ in the dual basis vectors $\hat{e}^a(x)$. The contravariant and covariant components of the vector field can be determined using equation (2.5):

$$\vec{v}(x) \cdot \hat{e}^\beta(x) = v^a(x)\hat{e}_a(x) \cdot \hat{e}^\beta(x) = v^a(x)\delta_a^\beta = v^\beta(x). \qquad (2.7)$$

Similarly,

$$v_a(x) = \vec{v}(x) \cdot \hat{e}_a(x). \qquad (2.8)$$

The coordinate basis vectors

In any particular coordinate system x^a, at every point P of the manifold with dimension N we can define a set of N coordinate basis vectors:

$$\hat{e}_\alpha = \lim_{\delta x^\alpha \to 0} \frac{\delta\vec{s}}{\delta x^\alpha}, \qquad (2.9)$$

where $\delta\vec{s}$ is the infinitesimal separation vector between point P and some nearby point Q with coordinate separation δx^α from P. Thus $\hat{e}_\alpha$ is the tangent vector to the x^α coordinate curve at the point P (figure 2.5).

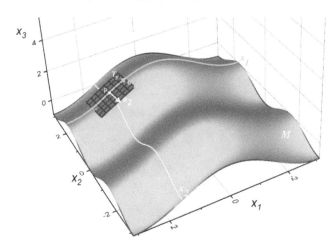

Figure 2.5. The basis vectors in the tangent space for a two-dimensional manifold.

Example 2.1. Find the coordinate basis vectors for a two-dimensional sphere embedded in a three-dimensional Euclidean space (see figure 2.6) defined by the following.
  (a) The Cartesian coordinates $(x, y) \to (x'^1, x'^2)$.
  (b) The spherical coordinates $(\theta, \varphi) \to (x^1, x^2)$.

  Solution:
  (a) Point P shown in figure 2.6 on a two-dimensional sphere can be described by the vector

$$\vec{s} = x'^1\hat{x} + x'^2\hat{y} \pm \sqrt{a^2 - (x'^1)^2 - (x'^2)^2}\,\hat{z}. \qquad (2.10)$$

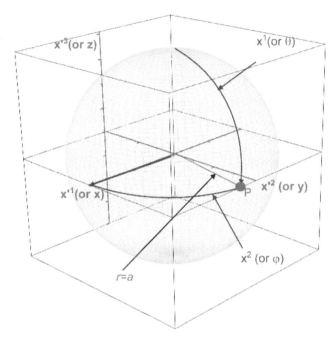

Figure 2.6. A two-dimensional sphere with radius $r = a$ embedded in three-dimensional Euclidean space. A point on the sphere is described by two coordinates (spherical $(\theta, \varphi) \to (x^1, x^2)$) and Cartesian $((x, y) \to (x'^1, x'^2))$.

Then one can write for the tangent vector at point P:

$$\delta\vec{s} = \delta x'^1 \hat{x} + \delta x'^2 \hat{y}$$

$$\mp \left(\frac{x'^1 \delta x'^1}{\sqrt{a^2 - (x'^1)^2 - (x'^2)^2}} + \frac{x'^2 \delta x'^2}{\sqrt{a^2 - (x'^1)^2 - (x'^2)^2}} \right) \hat{z}. \qquad (2.11)$$

The two basis vectors $\hat{e}_1(x'^1, x'^2)$ and $\hat{e}_2(x'^1, x'^2)$ at a point on the surface of the sphere are given by

$$\hat{e}_1 = \lim_{\delta x'^1 \to 0} \frac{\delta\vec{s}}{\delta x'^1}$$

$$= \lim_{\delta x'^1 \to 0} \left[\hat{x} + \frac{\delta x'^2}{\delta x'^1} \hat{y} \mp \left(\frac{x'^1}{\sqrt{a^2 - \left(x'^1\right)^2 - \left(x'^2\right)^2}} + \frac{x'^2 \frac{\delta x'^2}{\delta x'^1}}{\sqrt{a^2 - \left(x'^1\right)^2 - \left(x'^2\right)^2}} \right) \hat{z} \right],$$

$$\hat{e}_2 = \lim_{\delta x'^2 \to 0} \frac{\delta\vec{s}}{\delta x'^2} \qquad (2.12)$$

$$= \lim_{\delta x'^2 \to 0} \left[\frac{\delta x'^1}{\delta x'^2} \hat{x} + \hat{y} \mp \left(\frac{x'^1 \frac{\delta x'^1}{\delta x'^2}}{\sqrt{a^2 - \left(x'^1\right)^2 - \left(x'^2\right)^2}} + \frac{x'^2}{\sqrt{a^2 - \left(x'^1\right)^2 - \left(x'^2\right)^2}} \right) \hat{z} \right],$$

and noting that

$$\frac{\delta x^{'1}}{\delta x^{'2}} = \frac{\delta x^{'1}}{\delta x^{'2}} = 0, \tag{2.13}$$

we find

$$\hat{e}_1 = \hat{x} \mp \frac{x^{'1}}{\sqrt{a^2 - (x^{'1})^2 - (x^{'2})^2}} \hat{z},$$

$$\hat{e}_2 = \hat{y} \mp \frac{x^{'2}}{\sqrt{a^2 - (x^{'1})^2 - (x^{'2})^2}} \hat{z}. \tag{2.14}$$

(b) A point on a two-dimensional sphere in terms of the spherical coordinates
$(\theta, \varphi) \rightarrow (x^1, x^2)$ can be described by the vector

$$\vec{s} = a \sin(\theta) \cos(\varphi)\hat{x} + a \sin(\theta) \sin(\varphi)\hat{y} + a \cos(\theta)\hat{z}.$$
$$= a \sin(x^1) \cos(x^2)\hat{x} + a \sin(x^1) \sin(x^2)\hat{y} + a \cos(x^1)\hat{z}. \tag{2.15}$$

Then one can write

$$\delta\vec{s} = a \cos(x^1)\delta x^1[\cos(x^2)\hat{x} + \sin(x^2)\hat{y}] - a \sin(x^1)\delta x^1\hat{z}$$
$$+ a \sin(x^1)\delta x^2[\sin(x^2)\hat{x} - \cos(x^2)\hat{y}]. \tag{2.16}$$

The two basis vectors $\hat{e}_1(x^1, x^2)$ and $\hat{e}_2(x^1, x^2)$ at point P are given by

$$\hat{e}_1(x^1, x^2) = \lim_{\delta x^1 \to 0} \frac{\delta\vec{s}}{\delta x^1}$$
$$= \lim_{\delta x^1 \to 0} \left\{ a \cos(x^1)[\cos(x^2)\hat{x} + \sin(x^2)\hat{y}] - a \sin(x^1)\hat{z} \right.$$
$$\left. + a \sin(x^1)\frac{\delta x^2}{\delta x^1}[\sin(x^2)\hat{x} - \cos(x^2)\hat{y}] \right\},$$

$$\hat{e}_2(x^1, x^2) = \lim_{\delta x^2 \to 0} \frac{\delta\vec{s}}{\delta x^2}$$
$$= \lim_{\delta x^2 \to 0} \left\{ a \cos(x^1)\frac{\delta x^1}{\delta x^2}[\cos(x^2)\hat{x} + \sin(x^2)\hat{y}] - a \sin(x^1)\hat{z} \right.$$
$$\left. + a \sin(x^1)[\sin(x^2)\hat{x} - \cos(x^2)\hat{y}] \right\}, \tag{2.17}$$

and noting that

$$\frac{\delta x^1}{\delta x^2} = \frac{\delta x^1}{\delta x^2} = 0,$$

we find

$$\hat{e}_1(x^1, x^2) = a \cos (x^1)[\cos (x^2)\hat{x} + \sin (x^2)\hat{y}] - a \sin (x^1)\hat{z},$$
$$\hat{e}_2(x^1, x^2) = a \sin (x^1)[\sin (x^2)\hat{x} - \cos (x^2)\hat{y}]. \qquad (2.18)$$

Going back to the usual notations of the spherical coordinates $((\theta, \varphi) \rightarrow (x^1, x^2))$, this can be rewritten as

$$\hat{e}_1(\theta, \varphi) = a \cos (\theta)[\cos (\varphi)\hat{x} + \sin (\varphi)\hat{y}] - a \sin (\theta)\hat{z},$$
$$\hat{e}_2(\theta, \varphi) = a \sin (\theta)[\sin (\varphi)\hat{x} - \cos (\varphi)\hat{y}], \qquad (2.19)$$

so that one finds

$$\hat{e}_1(\theta, \varphi) = \cos (\theta) \cos (\varphi)\hat{x} + \cos (\theta) \sin (\varphi)\hat{y} - \sin (\theta)\hat{z} = \hat{e}_\theta,$$
$$\hat{e}_2(\theta, \varphi) = \sin (\theta) \sin (\varphi)\hat{x} - \sin (\theta) \cos (\varphi)\hat{y} = \hat{e}_\varphi, \qquad (2.20)$$

which are the usual unit vectors in spherical coordinates.

Example 2.2. Consider the two-dimensional manifold embedded in a three-dimensional Euclidean space shown in figure 2.7. Find the basis vectors $\hat{e}_1(x'^1, x'^2)$ and $\hat{e}_2(x'^1, x'^2)$.
Solution: We recall that this surface is defined by

$$x'^3(x'^1, x'^2) = \sin^2(x'^1) + \cos(x'^2), \qquad (2.21)$$

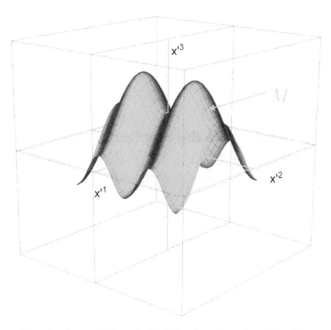

Figure 2.7. A two-dimensional manifold embedded in three-dimensional Euclidean space defined by $x^3(x'^1, x'^2) = \sin^2(x'^1) + \cos(x'^2)$.

where x'^1, x'^2, and x'^3 are the usual Cartesian coordinates (x, y, z). We note that a point on the surface can be defined by a vector

$$\vec{s} = x'^1\hat{x} + x'^2\hat{y} + \left(\sin^2(x'^1) + \cos(x'^2)\right)\hat{z} \tag{2.22}$$

so that

$$\delta\vec{s} = \delta x'^1\hat{x} + \delta x'^2\hat{y} + \left(2\sin(x'^1)\cos(x'^1)\delta x'^1 - \sin(x'^2)\delta x'^2\right)\hat{z}. \tag{2.23}$$

The two basis vectors $\hat{e}_1(x'^1, x'^2)$ and $\hat{e}_2(x'^1, x'^2)$ at point P in figure 2.7 are given by

$$\hat{e}_1(x'^1, x'^2) = \lim_{\delta x'^1 \to 0} \frac{\delta\vec{s}}{\delta x'^1}$$

$$= \lim_{\delta x'^1 \to 0}\left[\hat{x} + \frac{\delta x'^2}{\delta x'^1}\hat{y} + \left(2\sin(x'^1)\cos(x'^1) - \sin(x'^2)\frac{\delta x'^2}{\delta x'^1}\right)\hat{z}\right],$$

$$\hat{e}_2(x'^1, x'^2) = \lim_{\delta x'^2 \to 0} \frac{\delta\vec{s}}{\delta x'^2} \tag{2.24}$$

$$= \lim_{\delta x'^2 \to 0}\left[\frac{\delta x'^1}{\delta x'^2}\hat{x} + \hat{y} + \left(2\sin(x'^1)\cos(x'^1)\frac{\delta x'^1}{\delta x'^2} - \sin(x'^2)\right)\hat{z}\right],$$

and noting that

$$\frac{\delta x'^1}{\delta x'^2} = \frac{\delta x'^1}{\delta x'^2} = 0, \tag{2.25}$$

we find

$$\hat{e}_1\left(x'^1, x'^2\right) = \hat{x} + 2\sin\left(x'^1\right)\cos(x'^1)\hat{z},$$
$$\hat{e}_2\left(x'^1, x'^2\right) = \hat{y} - \sin(x'^2)\hat{z}. \tag{2.26}$$

2.4 The metric function and the coordinate basis vectors

Infinitesimal vector separation: Consider two points P and Q on a manifold with coordinates x^α and $x^\alpha + dx^\alpha$, where dx^α is non-zero for all α, then the infinitesimal vector separation between these two points is given by

$$d\vec{s} = \hat{e}_\alpha(x)dx^\alpha. \tag{2.27}$$

The metric function—the covariant components: The equation that determines the elements of the metric tensor in the metric

$$ds^2 = g_{\alpha\beta}(x)dx^\alpha dx^\beta \tag{2.28}$$

can be obtained from the inner product of the infinitesimal vector separation. We note that

$$ds^2 = d\vec{s} \cdot d\vec{s}$$
$$= \hat{e}_\alpha(x)dx^\alpha \cdot \hat{e}_\beta(x)dx^\beta = \hat{e}_\alpha(x) \cdot \hat{e}_\beta(x)dx^\alpha dx^\beta \qquad (2.29)$$
$$\Rightarrow ds^2 = g_{\alpha\beta}(x)dx^\alpha dx^\beta,$$

where

$$g_{\alpha\beta}(x) = \hat{e}_\alpha(x) \cdot \hat{e}_\beta(x) \qquad (2.30)$$

is the metric function (figures 2.8 and 2.9).

The metric function—the contravariant components: The contravariant components of the metric tensor can be defined as

$$g^{\alpha\beta}(x) = \hat{e}^\alpha(x) \cdot \hat{e}^\beta(x). \qquad (2.31)$$

Orthonormal basis vector: At a point on a manifold orthonormal basis vectors can be defined by

$$\hat{e}_\alpha(x) \cdot \hat{e}_\beta(x) = \eta_{\alpha\beta}, \qquad (2.32)$$

where

$$[\eta_{\alpha\beta}] = \begin{bmatrix} \pm 1 & 0 & \cdots & 0 \\ 0 & \pm 1 & \cdots & 0 \\ \vdots & & \ddots & \vdots \\ 0 & 0 & \cdots & \pm 1 \end{bmatrix}. \qquad (2.33)$$

Or, in short, $[\eta_{\alpha\beta}] = diag(\pm 1, \pm 1, \ldots, \pm 1)$ is the Cartesian line element of the tangent space T_p and depends on the signature of the pseudo-Riemannian manifold.

Example 2.3. Let us reconsider a two-dimensional sphere of radius a in three-dimensional Euclidean space. For an origin set at the north pole of the sphere, a point P is described by the vector (see figure 2.8)

$$\vec{s} = \rho \cos(\varphi)\hat{x} + \rho \sin(\varphi)\hat{y} + \sqrt{a^2 - \rho^2}\,\hat{z}$$
$$= x^1 \cos(x^2)\hat{x} + x^1 \sin(x^2)\hat{y} + \sqrt{a^2 - (x^1)^2}\,\hat{z}. \qquad (2.34)$$

(a) Find the basis vectors $\hat{e}_1(x^1, x^2)$ and $\hat{e}_2(x^1, x^2)$ in the tangent space at point P.
(b) Re-derive the metric elements for a two-dimensional sphere from the basis vectors.

Solution:
(a) We note that the tangent vector connecting point P with coordinate (x^1, x^2) with its neighboring point Q with coordinates $(x^1 + \delta x^1, x^2 + \delta x^2)$ is expressible as

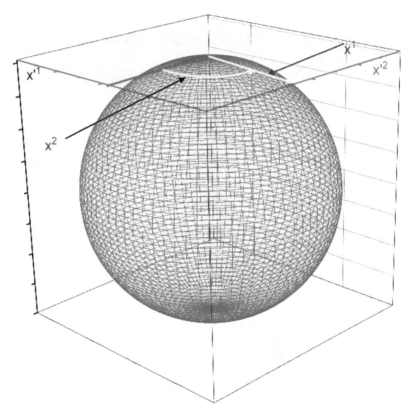

Figure 2.8. Two-dimensional sphere of radius a embedded in three-dimensional Euclidean space. The origin of the Cartesian coordinates (x'^1, x'^2, x'^3) (or (x, y, z)) is set at the north pole ($x^1 = \rho$, $x^2 = \varphi$).

$$\delta\vec{s} = \delta(x^1 \cos(x^2))\hat{x} + \delta(x^1 \sin(x^2))\hat{y} + \delta\left(\sqrt{a^2 - (x^1)^2}\right)\hat{z}$$

$$= \left[\cos(x^2)\hat{x} + \sin(x^2)\hat{y} - \frac{x^1}{\sqrt{a^2 - (x^1)^2}}\hat{z}\right]\delta x^1 \qquad (2.35)$$

$$- x^1[\sin(x^2)\hat{x} - \cos(x^2)\hat{y}]\delta x^2.$$

There follows that

$$\hat{e}_1 = \lim_{\delta x^1 \to 0} \frac{\delta\vec{s}}{\delta x^1}$$

$$= \cos(x^2)\hat{x} + \sin(x^2)\hat{y} - \frac{x^1}{\sqrt{a^2 - (x^1)^2}}\hat{z} \qquad (2.36)$$

$$- x^1[\sin(x^2)\hat{x} - \cos(x^2)\hat{y}]\frac{\delta x^2}{\delta x^1},$$

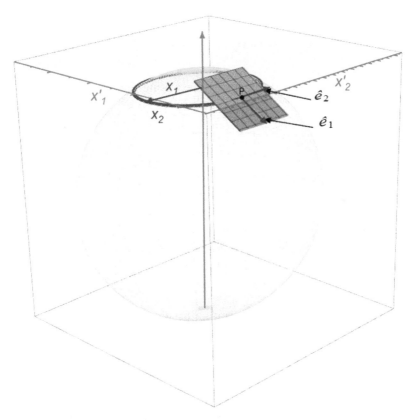

Figure 2.9. The basis vectors $(\hat{e}_1, \hat{e}_2)$ at a point P in the tangent space of a two-dimensional sphere embedded in three-dimensional Euclidean space.

and

$$\hat{e}_2 = \lim_{\delta x^2 \to 0} \frac{\delta \vec{s}}{\delta x^2}$$

$$= \left[\cos{(x^2)}\hat{x} + \sin{(x^2)}\hat{y} - \frac{x^1}{\sqrt{a^2 - (x^1)^2}}\hat{z} \right] \frac{\delta x^1}{\delta x^2} - x^1[\sin{(x^2)}\hat{x} - \cos{(x^2)}\hat{y}]. \tag{2.37}$$

Noting that

$$\frac{\delta x^1}{\delta x^2} = \frac{\delta x^2}{\delta x^1} = 0, \tag{2.38}$$

we find

$$\hat{e}_1 = \cos{(x^2)}\hat{x} + \sin{(x^2)}\hat{y} - \frac{x^1}{\sqrt{a^2 - (x^1)^2}}\hat{z} \tag{2.39}$$

and

$$\hat{e}_2 = -x^1[\sin(x^2)\hat{x} - \cos(x^2)\hat{y}]. \tag{2.40}$$

(see figure 2.9)

(b) The metric elements can be determined using the basis vectors

$$g_{\alpha\beta}(x) = \hat{e}_\alpha(x) \cdot \hat{e}_\beta(x), \tag{2.41}$$

and one finds

$$g_{11} = \hat{e}_1 \cdot \hat{e}_1 = \cos^2(x^2) + \sin^2(x^2) + \frac{(x^1)^2}{a^2 - (x^1)^2} = 1 + \frac{(x^1)^2}{a^2 - (x^1)^2}$$

$$\Rightarrow g_{11} = \frac{a^2}{a^2 - (x^1)^2}, \tag{2.42}$$

$$= \hat{e}_2 \cdot \hat{e}_2 = (x^1)^2[\sin^2(x^2) + \cos^2(x^2)] \Rightarrow g_{22} = (x^1)^2,$$

$$g_{12} = g_{21} = x^1[\cos(x^2)\hat{x} + \sin(x^2)\hat{y}] \cdot [-\sin(x^2)\hat{x} + \cos(x^2)\hat{y}]$$

$$g_{12} = g_{21} = 0.$$

These results are the same as found in chapter 1.

2.5 Basis vectors and coordinate transformations

Suppose we make a coordinate transformation from x^α where the basis vectors are $\hat{e}_\alpha$ to another set of coordinates x'^α with basis vectors $\hat{e}'_\alpha$. We want to find how the new basis vectors are transformed from $\hat{e}_\alpha$ to $\hat{e}'_\alpha$. To this end, we consider two points P and Q separated by an infinitesimal displacement $d\vec{s}$. This displacement is independent of the coordinate we chose to represent (figure 2.10). Therefore, we must have

$$d\vec{s} = \hat{e}_\alpha dx^\alpha = \hat{e}'_\mu dx'^\mu. \tag{2.43}$$

Using the coordinate transformation

$$dx^\alpha = \frac{\partial x^\alpha}{\partial x'^\beta} dx'^\beta, \tag{2.44}$$

one can write

$$\hat{e}_\alpha dx^\alpha = \frac{\partial x^\alpha}{\partial x'^\beta} \hat{e}_\alpha dx'^\beta = \hat{e}'_\mu dx'^\mu. \tag{2.45}$$

Upon substituting

$$\hat{e}'_\beta \cdot \hat{e}'^\beta = 1 \tag{2.46}$$

on the right side of equation (2.45), we can write

$$\frac{\partial x^\alpha}{\partial x'^\beta} \hat{e}_\alpha dx'^\beta = \left(\hat{e}'_\beta \cdot \hat{e}'^\beta\right)\hat{e}'_\mu dx'^\mu = \left(\hat{e}'_\mu \cdot \hat{e}'^\beta\right)\hat{e}'_\beta dx'^\mu = \hat{e}'_\beta \delta^\beta_\mu dx'^\mu \tag{2.47}$$

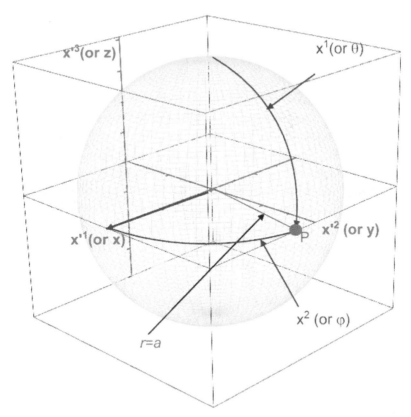

Figure 2.10. A two-dimensional sphere with radius $r = a$ embedded in three-dimensional Euclidean space. A point on the sphere is described by two coordinates (spherical $(\theta, \varphi) \to (x^1, x^2)$ and Cartesian $(x, y) \to (x'^1, x'^2)$).

$$\Rightarrow \frac{\partial x^\alpha}{\partial x'^\beta} \hat{e}_\alpha dx'^\beta = \hat{e}'_\beta dx'^\beta. \tag{2.48}$$

There follows that

$$\hat{e}'_\beta = \frac{\partial x^\alpha}{\partial x'^\beta} \hat{e}_\alpha. \tag{2.49}$$

For the dual basis vector, one can write

$$\hat{e}'_\alpha dx'^\alpha = \frac{\partial x'^\alpha}{\partial x^\beta} (dx^\beta \hat{e}'_\alpha) = \hat{e}_\alpha dx^\alpha$$

$$\Rightarrow \left(\frac{\partial x'^\alpha}{\partial x^\beta} \hat{e}^\eta \right) \cdot (dx^\beta \hat{e}'_\alpha) = \hat{e}^\eta \cdot \hat{e}_\alpha dx^\alpha = \delta^\eta_\alpha dx^\alpha \tag{2.50}$$

$$\Rightarrow \left(\frac{\partial x'^\alpha}{\partial x^\beta} \hat{e}^\eta \right) \cdot (\hat{e}'_\alpha dx^\beta) = \delta^\eta_\alpha dx^\alpha. \tag{2.51}$$

Replacing

$$\frac{\partial x'^{\alpha}}{\partial x^{\beta}}\hat{e}^{\eta} = \hat{e}'^{\alpha},$$

one can write

$$(\hat{e}'^{\alpha} \cdot \hat{e}'_{\alpha})(dx^{\beta}) = \delta^{\eta}_{\alpha}(dx^{\alpha}) \Rightarrow (\hat{e}'^{\alpha} \cdot \hat{e}'_{\alpha})(dx^{\beta}) = (dx^{\eta}), \tag{2.52}$$

where now

$$\frac{\partial x'^{\alpha}}{\partial x^{\beta}}\hat{e}^{\beta} = \hat{e}'^{\alpha},$$

since η is a dummy variable,

$$dx^{\beta} = dx^{\eta}, \tag{2.53}$$

when

$$\hat{e}'^{\alpha} = \frac{\partial x'^{\alpha}}{\partial x^{\beta}}\hat{e}^{\beta}. \tag{2.54}$$

Example 2.4. In example 2.1 we have shown that the basis vectors in the x^{α} and x'^{α} coordinates for a two-dimensional sphere (see figure 2.10) are given by

$$\hat{e}_1(x^1, x^2) = a \cos(x^1)[\cos(x^2)\hat{x} + \sin(x^2)\hat{y}] - a \sin(x^1)\hat{z},$$
$$\hat{e}_2(x^1, x^2) = a \sin(x^1)[\sin(x^2)\hat{x} - \cos(x^2)\hat{y}], \tag{2.55}$$

and

$$\hat{e}'_1 = \hat{x} \mp \frac{x'^1}{\sqrt{a^2 - (x'^1)^2 - (x'^2)^2}}\hat{z},$$
$$\hat{e}'_2 = \hat{y} \mp \frac{x'^2}{\sqrt{a^2 - (x'^1)^2 - (x'^2)^2}}\hat{z}. \tag{2.56}$$

Using the relation

$$\hat{e}_{\beta} = \frac{\partial x'^{\alpha}}{\partial x^{\beta}}\hat{e}'_{\alpha}, \tag{2.57}$$

show that

$$\hat{e}_1(x^1, x^2) = a \cos(x^1)[\cos(x^2)\hat{x} + \sin(x^2)\hat{y}] - a \sin(x^1)\hat{z}, \tag{2.58}$$

where the coordinate transformation is defined by

$$x'^1 = a \sin(x^1)\cos(x^2), \quad x'^2 = a \sin(x^1)\sin(x^2). \tag{2.59}$$

Solution: Noting that

$$\frac{\partial x'^1}{\partial x^1} = a\cos(x^1)\cos(x^2),\ \frac{\partial x'^1}{\partial x^2} = -a\sin(x^1)\sin(x^2)$$

$$\frac{\partial x'^2}{\partial x^1} = a\cos(x^1)\sin(x^2),\ \frac{\partial x'^2}{\partial x^2} = a\sin(x^1)\cos(x^2),$$

(2.60)

one can write

$$\hat{e}_1 = \frac{\partial x'^\alpha}{\partial x^1}\hat{e}'_\alpha = \frac{\partial x'^1}{\partial x^1}\hat{e}'_1 + \frac{\partial x'^2}{\partial x^1}\hat{e}'_2$$

$$= a\cos(x^1)\cos(x^2)\left(\hat{x} \mp \frac{x'^1}{\sqrt{a^2-(x'^1)^2-(x'^2)^2}}\hat{z}\right)$$

$$+ a\cos(x^1)\sin(x^2)\left(\hat{y} \mp \frac{x'^2}{\sqrt{a^2-(x'^1)^2-(x'^2)^2}}\hat{z}\right)$$

(2.61)

$$= a\cos(x^1)[\cos(x^2)\hat{x} + \sin(x^2)\hat{y}]$$

$$\mp a\left[\frac{x'^1\cos(x^1)\cos(x^2) + x'^2\cos(x^1)\sin(x^2)}{\sqrt{a^2-(x'^1)^2-(x'^2)^2}}\right]\hat{z}.$$

Noting that

$$x'^1 = a\sin(x^1)\cos(x^2),\ x'^2 = a\sin(x^1)\sin(x^2),$$

(2.62)

one can write

$$\frac{x'^1\cos(x^1)\cos(x^2) + x'^2\cos(x^1)\sin(x^2)}{\sqrt{a^2-(x'^1)^2-(x'^2)^2}} = \frac{a\sin(x^1)\cos(x^1)}{\sqrt{a^2-(x'^1)^2-(x'^2)^2}}.$$

(2.63)

We recall that for a two-dimensional sphere (in Cartesian coordinates)

$$x'^3 = \pm\sqrt{a^2-(x'^1)^2-(x'^2)^2} = a\cos(x^1),$$

(2.64)

so that

$$\pm\frac{x'^1\cos(x^1)\cos(x^2) + x'^2\cos(x^1)\sin(x^2)}{\sqrt{a^2-(x'^1)^2-(x'^2)^2}} = \frac{a\sin(x^1)\cos(x^1)}{\sqrt{a^2-(x'^1)^2-(x'^2)^2}}$$

(2.65)

$$= a\sin(x^1),$$

and the basis vector becomes

$$\hat{e}_1(x^1, x^2) = a\cos(x^1)[\cos(x^2)\hat{x} + \sin(x^2)\hat{y}] - a\sin(x^1)\hat{z}.$$

(2.66)

Example 2.5. Consider an observer (a man) on Earth (i.e., in the S frame) and another observer in the S' frame (Alien 2) that is moving with a speed v along the positive x-axis, as shown in figure 2.11. Suddenly, an event occurs: the appearance of Alien 1. This event is recorded by the two observers in the S and S' frames (i.e., the man and Alien 2). The man recorded this event at

$$(t, x, y, z) \to (ct, x, y, z) \to (x^0, x^1, x^2, x^3), \tag{2.67}$$

and Alien 2 recorded this event at

$$(t', x', y', z') \to (ct', x', y', z') \to \left(x'^0, x'^1, x'^2, x'^3\right). \tag{2.68}$$

These two coordinates are related by a Lorentz transformation for $x'^\alpha \to x^\alpha$:

$$x^0 = \gamma_v x'^0 + \gamma_v \beta x'^1, \ x^1 = \gamma_v \beta x'^0 + \gamma_v x'^1, \ x^2 = x'^2, \ x^3 = x'^3, \tag{2.69}$$

and for $x^\alpha \to x'^\alpha$:

$$x'^0 = \gamma_v x^0 - \gamma_v \beta x^1, \ x'^1 = \gamma_v \beta x^0 - \gamma_v x^1, \ x'^2 = x^2, \ x'^3 = x^3, \tag{2.70}$$

where

$$\beta = \frac{v}{c}, \ \gamma_v = \frac{1}{\sqrt{1 - \dfrac{v^2}{c^2}}}. \tag{2.71}$$

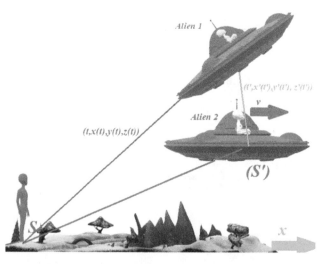

Figure 2.11. An event (appearance of Alien 1) recorded by two observers, an observer in the S frame (a man on Earth) and an observer in the S' frame (Alien 2 moving with a velocity v along the positive x-axis as measured by an observer in the S frame).

We recall that the metric for the Minkowski space-time described by the coordinates x^α is given by

$$(ds)^2 = g_{\alpha\beta}dx^\alpha dx^\beta = (dx^0)^2 - (dx^1)^2 - (dx^2)^2 - (dx^3)^2, \tag{2.72}$$

where the covariant components for the metric tensor are

$$[g_{\alpha\beta}] = \left[\vec{e}_\alpha(x) \cdot \vec{e}_\beta(x)\right] = [\eta_{\alpha\beta}] = \begin{bmatrix} 1 & 0 & 0 & 0 \\ 0 & -1 & 0 & 0 \\ 0 & 0 & -1 & 0 \\ 0 & 0 & 0 & -1 \end{bmatrix}. \tag{2.73}$$

(a) Find basis and dual basis vectors in the x'^α in terms of the basis vectors in the x^α:

$$\hat{e}'_\alpha = \Lambda'^\mu_\alpha \hat{e}_\mu, \quad \hat{e}'^\alpha = \Lambda'^\alpha_\mu \hat{e}^\mu,$$

where

$$\Lambda'^\mu_\alpha = \frac{\partial x^\mu}{\partial x'^\alpha}, \quad \Lambda'^\alpha_\mu = \frac{\partial x'^\alpha}{\partial x^\mu}.$$

(b) Using the basis and dual basis vectors in x'^α coordinates find the covariant and contravariant components for the metric tensor (i.e., find $g'_{\alpha\beta}$ and $g'^{\alpha\beta}$) in the x'^α coordinates.

Solution:
(a) Using the coordinate transformation

$$x'^0 = \gamma_v x^0 - \gamma_v\beta x^1, \ x'^1 = -\gamma_v\beta x^0 + \gamma_v x^1, \ x'^2 = x^2, \ x'^3 = x^3, \tag{2.74}$$

and the results from problem 3 in chapter 1, one can write

$$[\Lambda'^\alpha_\mu] = \left[\frac{\partial x'^\alpha}{\partial x^\mu}\right] = \begin{bmatrix} \gamma_v & -\gamma_v\beta & 0 & 0 \\ -\gamma_v\beta & \gamma_v & 0 & 0 \\ 0 & 0 & 1 & 0 \\ 0 & 0 & 0 & 1 \end{bmatrix}. \tag{2.75}$$

Noting that

$$[\Lambda'^\mu_\alpha] = [\Lambda'^\alpha_\mu]^T = \begin{bmatrix} \gamma_v & -\gamma_v\beta & 0 & 0 \\ -\gamma_v\beta & \gamma_v & 0 & 0 \\ 0 & 0 & 1 & 0 \\ 0 & 0 & 0 & 1 \end{bmatrix} = [\Lambda'^\alpha_\mu], \tag{2.76}$$

we find

$$[\hat{e}'_\alpha] = [\Lambda_\alpha^{'\mu}][\hat{e}_\mu] \Rightarrow \begin{bmatrix} \hat{e}'_0 \\ \hat{e}'_1 \\ \hat{e}'_2 \\ \hat{e}'_3 \end{bmatrix} = \begin{bmatrix} \gamma_v & -\gamma_v\beta & 0 & 0 \\ -\gamma_v\beta & \gamma_v & 0 & 0 \\ 0 & 0 & 1 & 0 \\ 0 & 0 & 0 & 1 \end{bmatrix} \begin{bmatrix} \hat{e}_0 \\ \hat{e}_1 \\ \hat{e}_2 \\ \hat{e}_3 \end{bmatrix} \tag{2.77}$$

$$\Rightarrow \hat{e}'_0 = \gamma_v\hat{e}_0 - \gamma_v\beta\hat{e}_1, \; \hat{e}'_1 = -\gamma_v\beta\hat{e}_0 + \gamma_v\hat{e}_1, \; \hat{e}'_2 = \hat{e}_2, \; \hat{e}'_3 = \hat{e}_3.$$

Since

$$[\Lambda_\alpha^{'\mu}] = [\Lambda_\mu^{'\alpha}]^{\mathrm{T}} = [\Lambda_\mu^{'\alpha}],$$

one can easily show that

$$\hat{e}'^\alpha = \Lambda_\mu^{'\alpha}\hat{e}^\mu \Rightarrow \begin{bmatrix} \hat{e}'^0 \\ \hat{e}'^1 \\ \hat{e}'^2 \\ \hat{e}'^3 \end{bmatrix} = \begin{bmatrix} \gamma_v & -\gamma_v\beta & 0 & 0 \\ -\gamma_v\beta & \gamma_v & 0 & 0 \\ 0 & 0 & 1 & 0 \\ 0 & 0 & 0 & 1 \end{bmatrix} \begin{bmatrix} \hat{e}^0 \\ \hat{e}^1 \\ \hat{e}^2 \\ \hat{e}^3 \end{bmatrix} \tag{2.78}$$

$$\Rightarrow \hat{e}'^0 = \gamma_v\hat{e}^0 - \gamma_v\beta\hat{e}^1, \; \hat{e}'_1 = -\gamma_v\beta\hat{e}^0 + \gamma_v\hat{e}^1, \; \hat{e}'^2 = \hat{e}^2, \; \hat{e}'^3 = \hat{e}^3.$$

(b) Noting that

$$g'_{\alpha\beta} = \hat{e}'_\alpha \cdot \hat{e}'_\beta = \Lambda_\alpha^{'\mu}\hat{e}_\mu \cdot \Lambda_\beta^{'\nu}\hat{e}_\nu = \Lambda_\alpha^{'\mu}\Lambda_\beta^{'\nu}(\hat{e}_\mu \cdot \hat{e}_\nu) = \Lambda_\alpha^{'\mu}\Lambda_\beta^{'\nu}g_{\mu\nu}$$

$$\Rightarrow g'_{\alpha\beta} = \Lambda_\alpha^{'\mu}g_{\mu\nu}\Lambda_\beta^{'\nu} \Rightarrow [g'_{\alpha\beta}] = [\Lambda_\alpha^{'\mu}][g_{\mu\nu}][\Lambda_\beta^{'\nu}]$$

$$\Rightarrow [g'_{\alpha\beta}] = \begin{bmatrix} \gamma_v & -\gamma_v\beta & 0 & 0 \\ -\gamma_v\beta & \gamma_v & 0 & 0 \\ 0 & 0 & 1 & 0 \\ 0 & 0 & 0 & 1 \end{bmatrix} \begin{bmatrix} 1 & 0 & 0 & 0 \\ 0 & -1 & 0 & 0 \\ 0 & 0 & -1 & 0 \\ 0 & 0 & 0 & -1 \end{bmatrix} \tag{2.79}$$

$$\times \begin{bmatrix} \gamma_v & -\gamma_v\beta & 0 & 0 \\ -\gamma_v\beta & \gamma_v & 0 & 0 \\ 0 & 0 & 1 & 0 \\ 0 & 0 & 0 & 1 \end{bmatrix},$$

which simplifies into

$$[g'_{\alpha\beta}] = \begin{bmatrix} \gamma_v & -\gamma_v\beta & 0 & 0 \\ -\gamma_v\beta & \gamma_v & 0 & 0 \\ 0 & 0 & 1 & 0 \\ 0 & 0 & 0 & 1 \end{bmatrix} \begin{bmatrix} \gamma_v & -\gamma_v\beta & 0 & 0 \\ \gamma_v\beta & -\gamma_v & 0 & 0 \\ 0 & 0 & -1 & 0 \\ 0 & 0 & 0 & -1 \end{bmatrix} \tag{2.80}$$

$$\Rightarrow \left[g'_{\alpha\beta} \right] = \begin{bmatrix} \gamma_v^2(1 - \beta^2) & 0 & 0 & 0 \\ 0 & -\gamma_v^2(1 - \beta^2) & 0 & 0 \\ 0 & 0 & -1 & 0 \\ 0 & 0 & 0 & -1 \end{bmatrix}. \tag{2.81}$$

Noting that

$$\beta = \frac{v}{c}, \gamma = \frac{1}{\sqrt{1 - \frac{v^2}{c^2}}} \Rightarrow \gamma_v^2(1 - \beta^2) = 1, \tag{2.82}$$

we find

$$\left[g'_{\alpha\beta} \right] = \begin{bmatrix} 1 & 0 & 0 & 0 \\ 0 & -1 & 0 & 0 \\ 0 & 0 & -1 & 0 \\ 0 & 0 & 0 & -1 \end{bmatrix} = [g_{\alpha\beta}]. \tag{2.83}$$

2.6 Components of a vector in coordinate transformations

In coordinate transformations the vector components are different but the vector itself is unchanged. Suppose the vector $\vec{v}$ is a vector at point P in the x^α coordinates and $\vec{v}'$ is a vector in the x'^α coordinates at the same point on the manifold. These vectors may be expressed in terms of the basis vectors in the two coordinates differently:

$$\vec{v} = v^\alpha \hat{e}_\alpha, \vec{v}' = v'^\alpha \hat{e}'_\alpha, \tag{2.84}$$

or in terms of the dual basis vectors:

$$\vec{v} = v_\alpha \hat{e}^\alpha, \vec{v}' = v'_\alpha \hat{e}'^\alpha. \tag{2.85}$$

However, the vector is the same since it describes a geometrical entity that is independent of the coordinate system. Therefore, we must have

$$\vec{v} = v^\alpha \hat{e}_\alpha = v'^\alpha \hat{e}'_\alpha, \tag{2.86}$$

so that taking the inner product of $\vec{v}$ and $\hat{e}'^\beta$, we can write

$$v'^\beta = \hat{e}'^\beta \cdot \vec{v} = v^\alpha \hat{e}'^\beta \cdot \hat{e}_\alpha, \tag{2.87}$$

where we used

$$\vec{v} = v^\alpha \hat{e}_\alpha = v'^\alpha \hat{e}'_\alpha. \tag{2.88}$$

Applying the relation in equation (2.54), one can write

$$\hat{e}'^{\beta} = \frac{\partial x'^{\beta}}{\partial x^{\mu}}\hat{e}^{\mu}, \tag{2.89}$$

so that

$$v'^{\beta} = v^{\alpha}\frac{\partial x'^{\beta}}{\partial x^{\mu}}\hat{e}^{\mu} \cdot \hat{e}_{\alpha} = v^{\alpha}\frac{\partial x'^{\beta}}{\partial x^{\mu}}\delta_{\alpha}^{\mu}, \tag{2.90}$$

which leads to

$$v'^{\beta} = \frac{\partial x'^{\beta}}{\partial x^{\alpha}}v^{\alpha}. \tag{2.91}$$

This shows how the components of a vector transform, which is similar to the basis vector transformation as one might have expected.

2.7 The inner product of vectors and the metric tensor

The inner product: At a point P on a manifold, the inner product of two vectors

$$\vec{v} = v^{\alpha}\hat{e}_{\alpha} \tag{2.92}$$

and

$$\vec{w} = w^{\beta}\hat{e}_{\beta} \tag{2.93}$$

is given by

$$\vec{v} \cdot \vec{w} = v^{\alpha}\hat{e}_{\alpha} \cdot w^{\beta}\hat{e}_{\beta} = g_{\alpha\beta}v^{\alpha}w^{\beta}, \tag{2.94}$$

where

$$g_{\alpha\beta} = \hat{e}_{\alpha} \cdot \hat{e}_{\beta} \tag{2.95}$$

are the covariant components of the metric tensor. When the vectors $\vec{v}(x)$ and $\vec{w}(x)$ are expressed in terms of dual basis vectors,

$$\vec{v} = v_{\alpha}\hat{e}^{\alpha}, \quad \vec{w} = w_{\beta}\hat{e}^{\beta}, \tag{2.96}$$

the inner product is given by

$$\vec{v} \cdot \vec{w} = v_{\alpha}\hat{e}^{\alpha} \cdot w_{\beta}\hat{e}^{\beta} = g^{\alpha\beta}v_{\alpha}w_{\beta}, \tag{2.97}$$

where

$$g^{\alpha\beta} = \hat{e}^{\alpha} \cdot \hat{e}^{\beta} \tag{2.98}$$

is the contravariant components of the metric tensor. We can also use the covariant and contravariant components of the vectors to determine the inner products:

$$\vec{v} \cdot \vec{w} = v^{\alpha}\hat{e}_{\alpha} \cdot w_{\beta}\hat{e}^{\beta} = \hat{e}_{\alpha} \cdot \hat{e}^{\beta}v^{\alpha}w_{\beta} = \delta_{\beta}^{\alpha}v^{\alpha}w_{\beta} = v^{\alpha}w_{\alpha}, \tag{2.99}$$

or

$$\vec{v} \cdot \vec{w} = v_\alpha \hat{e}^\alpha \cdot w^\beta \hat{e}_\beta = \hat{e}^\alpha \cdot \hat{e}_\beta v_\alpha w^\beta = \delta^\alpha_\beta v_\alpha w^\beta = v_\alpha w^\alpha. \tag{2.100}$$

Whichever way we determine the inner products, we must get the same values. Thus from equations (2.94) and (2.99), we find

$$\vec{v} \cdot \vec{w} = g_{\alpha\beta} v^\alpha w^\beta = v^\alpha w_\alpha \Rightarrow g_{\alpha\beta} w^\beta = w_\alpha. \tag{2.101}$$

Similarly, from equations (2.97) and (2.100), we find

$$\vec{v} \cdot \vec{w} = g^{\alpha\beta} v_\alpha w_\beta = v_\alpha w^\alpha \Rightarrow g^{\alpha\beta} w_\beta = w^\alpha. \tag{2.102}$$

From equation (2.101), we note that the covariant form of the metric tensor can be used to lower an index of a vector component. On the other hand, from equation (2.102), the contravariant form of the metric tensor can be used to raise an index of a vector component. Thus applying these properties of the metric tensor one can relate the basis and dual basis vectors by

$$\hat{e}_\alpha = g_{\alpha\nu} \hat{e}^\nu, \; \hat{e}^\mu = g^{\mu\beta} \hat{e}_\beta. \tag{2.103}$$

Then noting that the inner product

$$\hat{e}_\alpha \cdot \hat{e}^\mu = \delta^\alpha_\mu$$
$$\Rightarrow g_{\alpha\nu} \hat{e}^\nu \cdot g^{\mu\beta} \hat{e}_\beta = g_{\alpha\nu} g^{\mu\beta} \hat{e}^\nu \cdot \hat{e}_\beta = \delta^\alpha_\mu \tag{2.104}$$
$$\Rightarrow g_{\alpha\nu} g^{\mu\beta} \delta^\nu_\beta = \delta^\alpha_\mu \Rightarrow g_{\alpha\beta} g^{\mu\beta} = g^{\mu\beta} g_{\alpha\beta} = \delta^\alpha_\mu.$$

This means the metric tensor $[g^{\alpha\beta}]$ with the contravariant components $g^{\alpha\beta}$ is the inverse matrix of the metric tensor $[g_{\alpha\beta}]$ with the covariant elements $g_{\alpha\beta}$. Thus

$$[g^{\mu\beta}][g_{\alpha\beta}] = \left[\delta^\alpha_\mu \right] \Rightarrow G\tilde{G} = \tilde{G}G = I,$$

where $G = [g_{\alpha\beta}]$ is the metric tensor and $\tilde{G} = [g^{\alpha\beta}]$ is its inverse.

2.8 The inner product and the null vectors

Consider the inner product of two vectors $\vec{v}$ and $\vec{w}$ at a point on a manifold, which can be expressed in four different ways:

$$g_{\alpha\beta} v^\alpha w^\beta = v_\beta w^\beta = g^{\alpha\beta} v_\alpha w_\beta = v^\alpha w_\alpha. \tag{2.105}$$

Suppose $\vec{w} = \vec{v}$, equation (2.105) gives the inner product of vector $\vec{v}$ with itself:

$$g_{\alpha\beta} v^\alpha v^\beta = g^{\alpha\beta} v_\alpha v_\beta = v^\alpha v_\alpha = v_\alpha v^\alpha. \tag{2.106}$$

This result can be zero without the vector being actually a zero vector. We can see this if we recall the pseudo-Riemannian manifold metric:

$$(ds)^2 = g_{\alpha\beta}(x) dx^\alpha dx^\beta,$$

which could be zero or negative. To accommodate such vectors, we define the length of a vector $\vec{v}$ (v) as

$$v = \sqrt{\left| g_{\alpha\beta} v^\alpha v^\beta \right|} = \sqrt{\left| g^{\alpha\beta} v_\alpha v_\beta \right|} = \sqrt{\left| v^\alpha v_\alpha \right|} = \sqrt{\left| v_\beta v^\beta \right|}. \qquad (2.107)$$

As is the case in a pseudo-Riemannian manifold, the length of a vector can be zero without the vector being a zero vector (i.e., $v_\alpha \neq 0$). Zero vectors with non-zero components are known as *null vectors*.

The cosine of the angle between two vectors: The angle between two non-null vectors at a point on a manifold is defined by

$$\cos(\theta) = \frac{v^\alpha w_\alpha}{\sqrt{\left| v_\beta v^\beta \right|} \sqrt{\left| w_\mu w^\mu \right|}}. \qquad (2.108)$$

In the pseudo-Riemannian manifold, equation (2.108) can lead to $\left| \cos(\theta) \right| > 1$.

Orthogonal vectors: Two vectors,

$$\vec{v} = v^\alpha \hat{e}_\alpha, \;\; \vec{w} = w^\beta \hat{e}_\beta, \qquad (2.109)$$

are said to be orthogonal when the inner product of the two vectors is zero:

$$g_{\alpha\beta} v^\alpha w^\beta = g^{\alpha\beta} v_\alpha w_\beta = v^\alpha w_\alpha = v_\alpha w^\alpha = 0. \qquad (2.110)$$

2.9 The affine connections

It is important to know how vectors change as the coordinate or the parameter that defines the coordinate changes. For example, in Minkowski space-time, we may be interested in how the four-dimensional momentum $\vec{P}$ changes with the coordinates or the proper time (τ) that is used to parametrize the coordinates so that one can explain the condition for conservation of momentum in general relativity. This means, in four-dimensional momentum, which we may express in terms of its contravariant components,

$$\vec{P} = p^\alpha \hat{e}_\alpha, \qquad (2.111)$$

we may be interested in how the momentum changes with the coordinates, x^β,

$$\frac{\partial \vec{P}}{\partial x^\beta} = \frac{\partial}{\partial x^\beta}(p^\alpha \hat{e}_\alpha) = p^\alpha \frac{\partial \hat{e}_\alpha}{\partial x^\beta} + \hat{e}_\alpha \frac{\partial p^\alpha}{\partial x^\beta}, \qquad (2.112)$$

or how it changes with the proper time, τ,

$$\frac{d\vec{P}}{d\tau} = \frac{d}{d\tau}(p^\alpha \hat{e}_\alpha) = \hat{e}_\alpha \frac{dp^\alpha}{d\tau} + p^\alpha \frac{d\hat{e}_\alpha}{d\tau}, \qquad (2.113)$$

when the coordinates x^α are parametrized by the proper time $(x^\alpha(\tau))$. Equation (2.113), noting that the derivative of the basis vectors $\hat{e}_\alpha(x^\alpha(\tau))$ becomes

$$\frac{d\hat{e}_\alpha}{d\tau} = \frac{\partial \hat{e}_\alpha}{\partial x_\beta}\frac{dx^\beta}{d\tau},$$ (2.114)

can also be rewritten as

$$\frac{d\vec{P}}{d\tau} = \hat{e}_\alpha \frac{dp^\alpha}{d\tau} + p^\alpha \frac{dx^\beta}{d\tau}\frac{\partial \hat{e}_\alpha}{\partial x_\beta}.$$ (2.115)

Whether it is equation (2.112) or (2.115), we must know the derivative of the coordinate basis vector with respect to the coordinates or the parameter for the coordinates. That depends on what is known as the *affine connections*. To better understand affine connections, we shall reconsider the two-dimensional sphere embedded in a three-dimensional Euclidean space shown in figure 2.12. As one can see from the figure, the tangent space at point P is a plane with basis vectors defined by

$$\hat{e}_\rho(\rho, \varphi) = \cos(\varphi)\hat{x} + \sin(\varphi)\hat{y} - \frac{\rho}{\sqrt{a^2 - \rho^2}}\hat{z},$$

$$\hat{e}_\varphi(\rho, \varphi) = -\rho\sin(\varphi)\hat{x} + \rho\cos(\varphi)\hat{y},$$ (2.116)

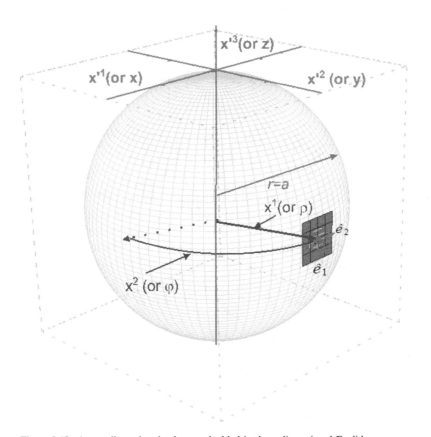

Figure 2.12. A two-dimensional sphere embedded in three-dimensional Euclidean space.

or, using (x^1, x^2) for (ρ, φ),

$$\hat{e}_1(x^1, x^2) = \cos(x^2)\hat{x} + \sin(x^2)\hat{y} - \frac{x^1}{\sqrt{a^2 - (x^1)^2}}\hat{z}. \tag{2.117}$$

$$\hat{e}_2(x^1, x^2) = -x^1 \sin(x^2)\hat{x} + x^1 \cos(x^2)\hat{y}.$$

Now, let us find the derivative of these basis vectors with respect to the coordinates. To this end, we note that

$$\frac{\partial \hat{e}_\rho(\rho, \varphi)}{\partial \varphi} = \frac{\partial \hat{e}_\varphi(\rho, \varphi)}{\partial \rho} = -\sin(\varphi)\hat{x} + \cos(\varphi)\hat{y} = \frac{1}{\rho}\hat{e}_\varphi$$

$$\Rightarrow \frac{\partial \hat{e}_\rho(\rho, \varphi)}{\partial \varphi} = \Gamma^2_{\rho\varphi}(\rho, \varphi)\hat{e}_\varphi, \quad \frac{\partial \hat{e}_\varphi(\rho, \varphi)}{\partial \rho} = \Gamma^2_{\varphi\rho}(\rho, \varphi)\hat{e}_\varphi, \tag{2.118}$$

where

$$\Gamma^2_{\rho\varphi}(\rho, \varphi) = \Gamma^2_{\varphi\rho}(\rho, \varphi) = \frac{1}{\rho} \tag{2.119}$$

is a function that connects the change in the basis vector $\hat{e}_\rho$ with respect to φ and that of $\hat{e}_\varphi$ with respect to ρ with the basis vector $\hat{e}_\varphi$. Note that in this particular case the change in the basis vectors results in basis vectors that are still in the same tangent space. Now let us consider

$$\frac{\partial \hat{e}_\rho(\rho, \varphi)}{\partial \rho} = -\left[\frac{1}{\sqrt{a^2 - \rho^2}} + \frac{\rho^2}{(a^2 - \rho^2)^{3/2}}\right]\hat{z} = -\frac{a^2}{(a^2 - \rho^2)^{3/2}}\hat{z} \tag{2.120}$$

and

$$\frac{\partial \hat{e}_\varphi(\rho, \varphi)}{\partial \varphi} = -\rho(\cos(\varphi)\hat{x} + \sin(\varphi)\hat{y}). \tag{2.121}$$

In this case, it is a little bit difficult to tell whether the change in the basis vector with respect to the change in the coordinates leads to a vector that is still in the same tangent space. So let us introduce a basis vector that is normal to the tangent space $\hat{e}_\perp$ (not part of the tangent space; see figure 2.13 for a closer look) in terms of the coordinate basis vectors:

$$\hat{e}_\perp(\rho, \varphi) = \hat{e}_\rho(\rho, \varphi) \times \hat{e}_\varphi(\rho, \varphi) = \frac{\rho^2}{\sqrt{a^2 - \rho^2}}(\cos(\varphi)\hat{x} + \sin(\varphi)\hat{y}) + \rho\hat{z}. \tag{2.122}$$

Combining this relation with the basis vector $\hat{e}_\rho(\rho, \varphi)$ in equation (2.116), one finds

$$\cos(\varphi)\hat{x} + \sin(\varphi)\hat{y} = \frac{1}{a}\sqrt{1 - \frac{\rho^2}{a^2}}\hat{e}_\perp - \frac{\rho^2}{a^2}\hat{e}_\rho$$

$$\hat{z} = \frac{a^2 - \rho^2}{a^2}\left(\frac{1}{\rho}\hat{e}_\perp - \frac{\rho}{\sqrt{a^2 - \rho^2}}\hat{e}_\rho\right). \tag{2.123}$$

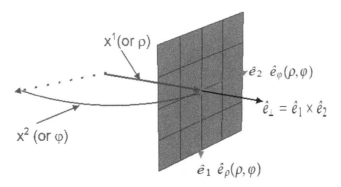

Figure 2.13. A vector normal $\hat{e}_\perp = \hat{e}_1 \times \hat{e}_2$ to the tangent space for point P on a two-dimensional sphere.

Upon substituting these equations into equations (2.120) and (2.121), we can then write

$$\frac{\partial \hat{e}_\rho(\rho, \varphi)}{\partial \rho} = -\frac{\rho}{\sqrt{a^2 - \rho^2}}\hat{e}_\rho + \frac{1}{\rho}\hat{e}_\perp, \tag{2.124}$$

$$\frac{\partial \hat{e}_\varphi(\rho, \varphi)}{\partial \varphi} = \frac{\rho^3}{a^2}\hat{e}_\rho - \frac{\rho}{a}\sqrt{1 - \frac{\rho^2}{a^2}}\hat{e}_\perp. \tag{2.125}$$

The results in equations (2.124) and (2.125) show that the change for the basis vectors with respect to the coordinates can lead to vectors with components that do not belong to the tangent space. Therefore, we must consider only the components that belong to the tangent space. In this particular case, one must then write

$$\frac{\partial \hat{e}_\rho(\rho, \varphi)}{\partial \rho} = \Gamma^1_{\rho\rho}(\rho, \varphi)\hat{e}_\rho, \quad \frac{\partial \hat{e}_\varphi(\rho, \varphi)}{\partial \varphi} = \Gamma^1_{\varphi\varphi}(\rho, \varphi)\hat{e}_\rho, \tag{2.126}$$

where

$$\Gamma^1_{\rho\rho}(\rho, \varphi) = -\frac{\rho}{\sqrt{a^2 - \rho^2}}, \; \Gamma^1_{\varphi\varphi}(\rho, \varphi) = \frac{\rho^3}{a^2}. \tag{2.127}$$

Then from equations (2.118) and (2.126), for a two-dimensional sphere embedded in a three-dimensional Euclidean space the derivative of the coordinate basis vectors can generally be written as

$$\frac{\partial \hat{e}_\rho(\rho, \varphi)}{\partial \rho} = \Gamma^1_{\rho\rho}(\rho, \varphi)\hat{e}_\rho + \Gamma^2_{\rho\rho}(\rho, \varphi)\hat{e}_\varphi,$$

$$\frac{\partial \hat{e}_\varphi(\rho, \varphi)}{\partial \varphi} = \Gamma^1_{\varphi\varphi}(\rho, \varphi)\hat{e}_\rho + \Gamma^2_{\varphi\varphi}(\rho, \varphi)\hat{e}_\varphi,$$

$$\frac{\partial \hat{e}_\varphi(\rho, \varphi)}{\partial \rho} = \Gamma^1_{\rho\varphi}(\rho, \varphi)\hat{e}_\rho + \Gamma^2_{\rho\varphi}(\rho, \varphi)\hat{e}_\varphi, \tag{2.128}$$

$$\frac{\partial \hat{e}_\rho(\rho, \varphi)}{\partial \varphi} = \Gamma^1_{\rho\varphi}(\rho, \varphi)\hat{e}_\rho + \Gamma^2_{\rho\varphi}(\rho, \varphi)\hat{e}_\varphi,$$

where

$$\Gamma^1_{\rho\rho}(\rho, \varphi) = -\frac{\rho}{\sqrt{a^2 - \rho^2}}, \Gamma^1_{\varphi\varphi}(\rho, \varphi) = \frac{\rho^3}{a^2},$$

$$\Gamma^2_{\rho\varphi}(\rho, \varphi) = \Gamma^2_{\varphi\rho}(\rho, \varphi) = \frac{1}{\rho},$$

$$\Gamma^2_{\rho\rho}(\rho, \varphi) = \Gamma^2_{\varphi\varphi}(\rho, \varphi) = \Gamma^1_{\rho\varphi}(\rho, \varphi) = \Gamma^1_{\varphi\rho}(\rho, \varphi) = 0.$$

(2.129)

In terms of the (x^1, x^2) for (ρ, φ), one can rewrite equation (2.128):

$$\frac{\partial \hat{e}_1(x^1, x^2)}{\partial x^2} = \Gamma^1_{12}(x^1, x^2)\hat{e}_1(x^1, x^2) + \Gamma^2_{12}(x^1, x^2)\hat{e}_2(x^1, x^2),$$

$$\frac{\partial \hat{e}_2(x^1, x^2)}{\partial x^1} = \Gamma^1_{21}(x^1, x^2)\hat{e}_1(x^1, x^2) + \Gamma^2_{21}(x^1, x^2)\hat{e}_2(x^1, x^2),$$

$$\frac{\partial \hat{e}_1(x^1, x^2)}{\partial x^1} = \Gamma^1_{11}(x^1, x^2)\hat{e}_1(x^1, x^2) + \Gamma^2_{11}(x^1, x^2)\hat{e}_2(x^1, x^2),$$

$$\frac{\partial \hat{e}_2(x^1, x^2)}{\partial x^2} = \Gamma^1_{22}(x^1, x^2)\hat{e}_1(x^1, x^2) + \Gamma^2_{22}(x^1, x^2)\hat{e}_2(x^1, x^2),$$

(2.130)

where

$$\Gamma^1_{11}(x^1, x^2) = -\frac{x^1}{\sqrt{a^2 - (x^1)^2}}, \Gamma^1_{22}(x^1, x^2) = \frac{(x^1)^3}{a^2},$$

$$\Gamma^2_{12}(x^1, x^2) = \Gamma^2_{21}(x^1, x^2) = \frac{1}{x^1},$$

$$\Gamma^2_{11}(x^1, x^2) = \Gamma^2_{22}(x^1, x^2) = \Gamma^1_{12}(x^1, x^2) = \Gamma^1_{21}(x^1, x^2) = 0.$$

(2.131)

Note that $\Gamma^\beta_{\alpha\mu}(x^1, x^2)$ (for $\alpha, \beta, \mu = 1, 2)$) represents the coefficients resulting from the derivative of the basis vectors that are only in the tangent space at point P, (i.e., components parallel to the basis vector $\hat{e}_\beta(x)$). Furthermore, we note that for a two-dimensional manifold there are $8 = 2^3$ functions $\Gamma^\beta_{\alpha\mu}(x^1, x^2)$.

Generally, for non-spherical geometry or a non-Euclidean manifold, the change in the basis vectors with respect to the coordinate at a point on a manifold (derivative of the basis vector $\hat{e}_\alpha$ with respect to the coordinate x^μ at that point) can have components inside or outside the tangent space. However, since we are confined to the tangent space, for those components outside the tangent space we consider only its projection that belongs to the tangent space at point P (parallel to the tangent space):

$$\frac{\partial \hat{e}_\alpha}{\partial x^\mu} = \left(\lim_{\delta x^\mu \to 0} \frac{\delta \hat{e}_\alpha}{\delta x^\mu}\right)_{\|T_p} = \Gamma^1_{\alpha\mu}\hat{e}_1 + \Gamma^2_{\alpha\mu}\hat{e}_2 + \Gamma^3_{\alpha\mu}\hat{e}_3 \cdots + \Gamma^N_{\alpha\mu}\hat{e}_N$$

$$\Rightarrow \frac{\partial \hat{e}_\alpha}{\partial x^\mu} = \Gamma^\beta_{\alpha\mu}(x)\hat{e}_\beta(x),$$

(2.132)

where N is the dimension of the tangent space. The N^3 coefficients $\Gamma^\beta_{\alpha\mu}(x)$ are known collectively as *the affine connection* or the *Christoffel symbol of the second kind* at point P.

Taking the inner product of equation (2.132) and the dual basis vectors,

$$\hat{e}^\nu \cdot \frac{\partial \hat{e}_\alpha}{\partial x^\mu} = \Gamma^\beta_{\alpha\mu} \hat{e}^\nu \cdot \hat{e}_\beta, \tag{2.133}$$

so that employing the relation

$$\hat{e}^\nu \cdot \hat{e}_\beta = \delta^\nu_\beta, \tag{2.134}$$

we find

$$\hat{e}^\nu \cdot \frac{\partial \hat{e}_\alpha}{\partial x^\mu} = \Gamma^\beta_{\alpha\mu} \delta^\nu_\beta \Rightarrow \Gamma^\nu_{\alpha\mu} = \hat{e}^\nu \cdot \frac{\partial \hat{e}_\alpha}{\partial x^\mu}, \tag{2.135}$$

which can be rewritten as

$$\Gamma^\beta_{\alpha\mu} = \hat{e}^\beta \cdot \frac{\partial \hat{e}_\alpha}{\partial x^\mu} = \hat{e}^\beta \cdot \partial_\mu \hat{e}_\alpha, \tag{2.136}$$

where we changed ν to β and introduced the notation

$$\partial_\mu \hat{e}_\alpha = \frac{\partial \hat{e}_\alpha}{\partial x^\mu}. \tag{2.137}$$

Upon differentiating

$$\hat{e}^\alpha \cdot \hat{e}_\beta = \delta^\alpha_\beta, \tag{2.138}$$

with respect to the coordinate x^μ, one can write

$$\begin{aligned}
\hat{e}^\alpha \cdot \partial_\mu \hat{e}_\beta + \hat{e}_\beta \cdot \partial_\mu \hat{e}^\alpha &= \partial_\mu \delta^\beta_\alpha = 0 \\
\Rightarrow \hat{e}_\beta \cdot \partial_\mu \hat{e}^\alpha &= -\hat{e}^\alpha \cdot \partial_\mu \hat{e}_\beta.
\end{aligned} \tag{2.139}$$

Applying the relation in equation (2.136) to equation (2.139), one can write

$$\hat{e}_\beta \cdot \partial_\mu \hat{e}^\alpha = -\hat{e}^\alpha \cdot \partial_\mu \hat{e}_\beta = -\Gamma^\alpha_{\beta\mu}. \tag{2.140}$$

Upon substituting

$$\hat{e}_\beta \cdot \hat{e}^\beta = 1, \tag{2.141}$$

into equation (2.140), we find

$$\begin{aligned}
\hat{e}_\beta \cdot \partial_\mu \hat{e}^\alpha &= -\Gamma^\alpha_{\beta\mu} \hat{e}_\beta \cdot \hat{e}^\beta = \hat{e}_\beta \cdot \left(-\Gamma^\alpha_{\beta\mu} \hat{e}^\beta \right) \\
\Rightarrow \partial_\mu \hat{e}^\alpha &= -\Gamma^\alpha_{\beta\mu} \hat{e}^\beta.
\end{aligned} \tag{2.142}$$

Note that the derivative of the dual basis vectors with respect to the coordinates has the same form except the minus sign.

2.10 The affine connection under coordinate transformation

Suppose we make the coordinate transformation $x^\alpha \to x'^\alpha$. The affine connections in these two coordinates are given by

$$\Gamma^\beta_{\alpha\mu} = \hat{e}^\beta \cdot \frac{\partial \hat{e}_\alpha}{\partial x^\mu}, \ \Gamma'^\beta_{\alpha\mu} = \hat{e}'^\beta \cdot \frac{\partial \hat{e}'_\alpha}{\partial x'^\mu}. \tag{2.143}$$

In this section, we want to find the equation that relates the affine connection $\Gamma'^\beta_{\alpha\mu}$ to the affine connection $\Gamma^\beta_{\alpha\mu}$. To this end, applying the relations for the basis vectors for such coordinate transformation,

$$\hat{e}'_\alpha = \frac{\partial x^\sigma}{\partial x'^\alpha}\hat{e}_\sigma, \ \hat{e}'^\beta = \frac{\partial x'^\beta}{\partial x^\nu}\hat{e}^\nu, \tag{2.144}$$

one can rewrite $\Gamma'^\beta_{\alpha\mu}$ as

$$\begin{aligned}
\Gamma'^\beta_{\alpha\mu} &= \hat{e}'^\beta \cdot \frac{\partial \hat{e}'_\alpha}{\partial x'^\mu} = \frac{\partial x'^\beta}{\partial x^\nu}\hat{e}^\nu \cdot \frac{\partial}{\partial x'^\mu}\left(\frac{\partial x^\sigma}{\partial x'^\alpha}\hat{e}_\sigma\right) \\
&= \frac{\partial x'^\beta}{\partial x^\nu}\hat{e}^\nu \cdot \left[\frac{\partial \hat{e}_\sigma}{\partial x'^\mu}\frac{\partial x^\sigma}{\partial x'^\alpha} + \hat{e}_\sigma\frac{\partial^2 x^\sigma}{\partial x'^\mu \partial x'^\alpha}\right] \\
&= \frac{\partial x'^\beta}{\partial x^\nu}\frac{\partial x^\sigma}{\partial x'^\alpha}\hat{e}^\nu \cdot \frac{\partial \hat{e}_\sigma}{\partial x'^\mu} + \frac{\partial^2 x^\sigma}{\partial x'^\mu \partial x'^\alpha}\frac{\partial x'^\beta}{\partial x^\nu}\hat{e}^\nu \cdot \hat{e}_\sigma.
\end{aligned} \tag{2.145}$$

Employing the chain rule, one can write

$$\frac{\partial \hat{e}_\sigma}{\partial x'^\mu} = \frac{\partial \hat{e}_\sigma}{\partial x^g}\frac{\partial x^\kappa}{\partial x'^\mu}, \tag{2.146}$$

so that equation (2.145) becomes

$$\Gamma'^\beta_{\alpha\mu} = \frac{\partial x'^\beta}{\partial x^\nu}\frac{\partial x^\sigma}{\partial x'^\alpha}\frac{\partial x^\kappa}{\partial x'^\mu}\hat{e}^\nu \cdot \frac{\partial \hat{e}_\sigma}{\partial x^\kappa} + \frac{\partial^2 x^\sigma}{\partial x'^\mu \partial x'^\alpha}\frac{\partial x'^\beta}{\partial x^\nu}\hat{e}^\nu \cdot \hat{e}_\sigma, \tag{2.147}$$

and substituting

$$\hat{e}^\nu \cdot \frac{\partial \hat{e}_\sigma}{\partial x^\kappa} = \Gamma^\nu_{\sigma\kappa}, \ \hat{e}^\nu \cdot \hat{e}_\sigma = \delta^\sigma_\nu, \tag{2.148}$$

equation (2.147) leads to

$$\Gamma'^\beta_{\alpha\mu} = \frac{\partial x'^\beta}{\partial x^\nu}\frac{\partial x^\sigma}{\partial x'^\alpha}\frac{\partial x^\kappa}{\partial x'^\mu}\Gamma^\nu_{\sigma\kappa} + \frac{\partial x'^\beta}{\partial x^\nu}\frac{\partial^2 x^\nu}{\partial x'^\mu \partial x'^\alpha}, \tag{2.149}$$

which is the equation that transforms the affine connection in the x^α coordinates to the x'^α coordinates. This can also be rewritten as

$$\Gamma'^\beta_{\alpha\mu} = \frac{\partial x'^\beta}{\partial x^\nu}\frac{\partial x^\sigma}{\partial x'^\alpha}\frac{\partial x^\kappa}{\partial x'^\mu}\Gamma^\nu_{\sigma\kappa} - \frac{\partial x^\nu}{\partial x'^\alpha}\frac{\partial x^\sigma}{\partial x'^\mu}\frac{\partial^2 x'^\beta}{\partial x^\nu \partial x^\sigma}. \tag{2.150}$$

(See problem 4).

2.11 The affine connection and the metric tensor

Next, we shall derive the equation that relates the affine connections to the metric tensor. To this end, we recall the affine connection

$$\Gamma^{\beta}_{\alpha\mu} = \hat{e}^{\beta} \cdot \frac{\partial \hat{e}_{\alpha}}{\partial x^{\mu}}. \tag{2.151}$$

In a similar manner, by interchanging the indices of the base vector and the coordinate, we can also write

$$\Gamma^{\beta}_{\mu\alpha} = \hat{e}^{\beta} \cdot \frac{\partial \hat{e}_{\mu}}{\partial x^{\alpha}}. \tag{2.152}$$

The difference between equations (2.151) and (2.152), which we represent by $T^{\beta}_{\alpha\mu}$,

$$T^{\beta}_{\alpha\mu} = \Gamma^{\beta}_{\alpha\mu} - \Gamma^{\beta}_{\mu\alpha}, \tag{2.153}$$

is known as the *torsion tensor*. For a *torsionless* manifold,

$$T^{\beta}_{\alpha\mu} = 0 \Rightarrow \Gamma^{\beta}_{\alpha\mu} = \Gamma^{\beta}_{\mu\alpha}. \tag{2.154}$$

We will determine the relationship between the affine connection and the metric tensor for a torsionless manifold. We recall the metric tensor

$$g_{\alpha\beta} = \hat{e}_{\alpha} \cdot \hat{e}_{\beta}, \tag{2.155}$$

so that

$$\frac{\partial g_{\alpha\beta}}{\partial x^{\mu}} = \partial_{\mu} g_{\alpha\beta} = \partial_{\mu}\left(\hat{e}_{\alpha} \cdot \hat{e}_{\beta}\right) = \hat{e}_{\alpha} \cdot \partial_{\mu}\hat{e}_{\beta} + \hat{e}_{\beta} \cdot \partial_{\mu}\hat{e}_{\alpha}. \tag{2.156}$$

Applying the relation

$$\partial_{\mu}\hat{e}_{\beta} = \Gamma^{\nu}_{\beta\mu}\hat{e}_{\nu}, \quad \partial_{\mu}\hat{e}_{\alpha} = \Gamma^{\nu}_{\alpha\mu}\hat{e}_{\nu}, \tag{2.157}$$

we can rewrite equation (2.156) as

$$\partial_{\mu} g_{\alpha\beta} = \Gamma^{\nu}_{\alpha\mu}\hat{e}_{\beta} \cdot \hat{e}_{\nu} + \Gamma^{\nu}_{\beta\mu}\hat{e}_{\alpha} \cdot \hat{e}_{\nu} = \Gamma^{\nu}_{\alpha\mu}g_{\beta\nu} + \Gamma^{\nu}_{\beta\mu}g_{\alpha\nu}. \tag{2.158}$$

Similarly,

$$\begin{aligned} \partial_{\beta} g_{\mu\alpha} &= \partial_{\beta}\left(\hat{e}_{\mu} \cdot \hat{e}_{\alpha}\right) = \hat{e}_{\mu} \cdot \partial_{\beta}\hat{e}_{\alpha} + \hat{e}_{\alpha} \cdot \partial_{\beta}\hat{e}_{\mu} \\ &= \Gamma^{\nu}_{\alpha\beta}\hat{e}_{\mu} \cdot \hat{e}_{\nu} + \Gamma^{\nu}_{\mu\beta}\hat{e}_{\alpha} \cdot \hat{e}_{\nu} \\ &\Rightarrow \partial_{\beta} g_{\mu\alpha} = \Gamma^{\nu}_{\alpha\beta}g_{\mu\nu} + \Gamma^{\nu}_{\mu\beta}g_{\alpha\nu}, \end{aligned} \tag{2.159}$$

and

$$\partial_{\alpha} g_{\beta\mu} = \Gamma^{\nu}_{\beta\alpha}g_{\nu\mu} + \Gamma^{\nu}_{\mu\alpha}g_{\beta\nu}. \tag{2.160}$$

Now, combining equations (2.158)–(2.160), we can write

$$\partial_\mu g_{\alpha\beta} + \partial_\beta g_{\mu\alpha} - \partial_\alpha g_{\beta\mu} =$$
$$\Gamma^\nu_{\alpha\mu} g_{\beta\nu} + \Gamma^\nu_{\beta\mu} g_{\alpha\nu} + \Gamma^\nu_{\alpha\beta} g_{\mu\nu} + \Gamma^\nu_{\mu\beta} g_{\alpha\nu} - \Gamma^\nu_{\beta\alpha} g_{\nu\mu} - \Gamma^\nu_{\mu\alpha} g_{\beta\nu}. \tag{2.161}$$

Recalling that we are interested in a torsionless manifold, where $\Gamma^\beta_{\alpha\mu} = \Gamma^\beta_{\mu\alpha}$, equation (2.161) reduces to

$$\partial_\mu g_{\alpha\beta} + \partial_\beta g_{\mu\alpha} - \partial_\alpha g_{\beta\mu} = 2\Gamma^\nu_{\beta\mu} g_{\alpha\nu}. \tag{2.162}$$

Multiplying equation (2.163) by $g^{\eta\alpha}$,

$$g^{\eta\alpha}\left(\partial_\mu g_{\alpha\beta} + \partial_\beta g_{\mu\alpha} - \partial_\alpha g_{\beta\mu}\right) = 2\Gamma^\nu_{\beta\mu} g^{\eta\alpha} g_{\alpha\nu}. \tag{2.163}$$

Recalling that

$$g^{\eta\alpha} g_{\alpha\nu} = \delta^\eta_\nu,$$

equation (2.104) can be put in the form

$$g^{\eta\alpha}\left(\partial_\mu g_{\alpha\beta} + \partial_\beta g_{\mu\alpha} - \partial_\alpha g_{\beta\mu}\right) = 2\Gamma^\nu_{\beta\mu} \delta^\eta_\nu$$
$$\Rightarrow \Gamma^\eta_{\beta\mu} = \frac{g^{\eta\alpha}}{2}\left(\partial_\mu g_{\alpha\beta} + \partial_\beta g_{\mu\alpha} - \partial_\alpha g_{\beta\mu}\right). \tag{2.164}$$

Now, relabeling the index α by ν:

$$g^{\eta\nu}\left(\partial_\mu g_{\nu\beta} + \partial_\beta g_{\mu\nu} - \partial_\nu g_{\beta\mu}\right) = 2\Gamma^\eta_{\beta\mu}, \tag{2.165}$$

and then η by α, we can write

$$\Gamma^\alpha_{\beta\mu} = \frac{g^{\alpha\nu}}{2}\left(\partial_\beta g_{\mu\nu} + \partial_\mu g_{\nu\beta} - \partial_\nu g_{\beta\mu}\right). \tag{2.166}$$

The right-hand side of equation (2.166) is known as the *metric connection* and is often represented by $\left\{ {\alpha \atop \beta\mu} \right\}$.

We can use the raising and lowering properties of the metric tensor to derive some important formulas involving the affine connections and the metric tensor. These include the following:

$$\Gamma_{\alpha\beta\mu} = g_{\alpha\nu} \Gamma^\nu_{\beta\mu}. \tag{2.167}$$

Multiplying by $g^{\sigma\alpha}$,

$$g^{\sigma\alpha} \Gamma_{\alpha\beta\mu} = g^{\sigma\alpha} g_{\alpha\nu} \Gamma^\nu_{\beta\mu} = \delta^\sigma_\nu \Gamma^\nu_{\beta\mu} = \Gamma^\sigma_{\beta\mu}$$
$$\Rightarrow \Gamma^\sigma_{\beta\mu} = g^{\sigma\alpha} \Gamma_{\alpha\beta\mu}. \tag{2.168}$$

Applying the relation in equation (2.166), we can express equation (2.167) as

$$\Gamma^\sigma_{\beta\mu} = \frac{g^{\sigma\nu}}{2}\left(\partial_\beta g_{\mu\nu} + \partial_\mu g_{\nu\beta} - \partial_\nu g_{\beta\mu}\right) = g^{\sigma\alpha} \Gamma_{\alpha\beta\mu}. \tag{2.169}$$

In equation (2.169), ν is a summation index and can be replaced by α,

$$\Gamma^{\sigma}_{\beta\mu} = \frac{g^{\sigma\alpha}}{2}\left(\partial_{\beta}g_{\mu\alpha} + \partial_{\mu}g_{\alpha\beta} - \partial_{\alpha}g_{\beta\mu}\right) = g^{\sigma\alpha}\Gamma_{\alpha\beta\mu}, \tag{2.170}$$

so that one finds

$$\Gamma_{\alpha\beta\mu} = \frac{1}{2}\left(\partial_{\beta}g_{\mu\alpha} + \partial_{\mu}g_{\alpha\beta} - \partial_{\alpha}g_{\beta\mu}\right). \tag{2.171}$$

The quantity $\Gamma_{\alpha\beta\mu}$ is traditionally known as a *Christoffel symbol of the first kind.* Noting in view of equation (2.171), by switching places for the index α and β, one can write

$$\Gamma_{\beta\alpha\mu} = \frac{1}{2}\left(\partial_{\alpha}g_{\mu\beta} + \partial_{\mu}g_{\beta\alpha} - \partial_{\beta}g_{\alpha\mu}\right), \tag{2.172}$$

so that

$$\Gamma_{\alpha\beta\mu} + \Gamma_{\beta\alpha\mu}$$
$$= \frac{1}{2}\left(\partial_{\beta}g_{\mu\alpha} + \partial_{\mu}g_{\alpha\beta} - \partial_{\alpha}g_{\beta\mu} + \partial_{\alpha}g_{\mu\beta} + \partial_{\mu}g_{\beta\alpha} - \partial_{\beta}g_{\alpha\mu}\right), \tag{2.173}$$

and taking into account the symmetry of the metric tensor

$$g_{\mu\alpha} = g_{\alpha\mu}, \, g_{\alpha\beta} = g_{\beta\alpha}, \, g_{\beta\mu} = g_{\mu\beta},$$

from equation (2.173), we find

$$\partial_{\mu}g_{\alpha\beta} = \Gamma_{\alpha\beta\mu} + \Gamma_{\beta\alpha\mu}. \tag{2.174}$$

Equation (2.174) allows us to express the partial derivative of the metric components in terms of the affine connection.

We recall the determinant of any $n \times n$ square matrix A is given by

$$\det |A| = \sum_{j=1}^{n}(-1)^{i+j}a_{ij}M_{ij}, \tag{2.175}$$

where $i = 1, \, 2, \, 3 \ldots$ or n, and M_{ij} is called the *minor* of a_{ij}, which is the determinant of the matrix constructed by removing the row and the column containing a_{ij} (i.e., the ith row and the jth column). This can also be expressed in terms of the cofactor matrix for A, as

$$\det |A| = \sum_{j=1}^{n}a_{ij}[cof(G)]_i^j = a_{ij}[cof(G)]_i^j, \tag{2.176}$$

where

$$[cof(G)]_i^j = (-1)^{i+j}M_{ij}. \tag{2.177}$$

In view of this relation, one can write the determinant of the metric tensor:

$$g = \det |G| = g_{\alpha\beta}[cof(G)]_\alpha^\beta. \tag{2.178}$$

As an example, let us consider a two-dimensional metric tensor:

$$[g_{\alpha\beta}] = \begin{bmatrix} g_{11} & g_{12} \\ g_{21} & g_{22} \end{bmatrix}.$$

According to equation (2.178), the determinant is given by

$$g = \det |G| = g_{\alpha\beta}[cof(G)]_\alpha^\beta = g_{11}[cof(G)]_1^1 + g_{12}[cof(G)]_1^2, \tag{2.179}$$

or

$$g = \det |G| = g_{\alpha\beta}[cof(G)]_\alpha^\beta = g_{21}[cof(G)]_2^1 + g_{22}[cof(G)]_2^2. \tag{2.180}$$

Now, using the cofactor matrix

$$[cof(G)] = \begin{bmatrix} g_{22} & -g_{21} \\ -g_{12} & g_{11} \end{bmatrix},$$

we find

$$g = \det |G| = g_{11}g_{22} - g_{21}g_{12} \tag{2.181}$$

or

$$g = \det |G| = -g_{21}g_{21} + g_{11}g_{22} = g_{11}g_{22} - g_{21}g_{12}, \tag{2.182}$$

which are the same. Now if we take the derivative of this result with respect to the coordinate x^μ, we have

$$\frac{\partial g}{\partial x^\mu} = \det |G| = \frac{\partial g_{11}}{\partial x^\mu}[cof(G)]_1^1 + g_{11}\frac{\partial}{\partial x^\mu}[cof(G)]_1^1$$

$$+ \frac{\partial g_{12}}{\partial x^\mu}[cof(G)]_1^2 + g_{12}\frac{\partial}{\partial x^\mu}[cof(G)]_1^2, \tag{2.183}$$

$$= (g_{22})\frac{\partial g_{11}}{\partial x^\mu} + (g_{11})\frac{\partial g_{22}}{\partial x^\mu} + (-g_{12})\frac{\partial g_{21}}{\partial x^\mu}(-g_{21})\frac{\partial g_{12}}{\partial x^\mu},$$

so that one can write

$$\frac{\partial g}{\partial x^\mu} = [cof(G)]^{\alpha\beta}\partial_\mu g_{\alpha\beta}, \tag{2.184}$$

for $\alpha, \beta = 1, 2$. This clearly shows the change in determinant with respect to coordinates can be expressed in terms of only the change in elements in the metric tensor with respect to the coordinates:

$$\partial_\mu(g) = [cof(G)]^{\alpha\beta}\partial_\mu g_{\alpha\beta}. \tag{2.185}$$

Recalling that for the metric tensor

$$[g^{\alpha\beta}] = [g_{\alpha\beta}]^{-1}, \tag{2.186}$$

and the inverse of a square matrix G is given by

$$G^{-1} = \frac{[cof(G)]^{\mathrm{T}}}{\det |G|} \Rightarrow [cof(G)]^{\mathrm{T}} = G^{-1}\det |G|, \tag{2.187}$$

one can write

$$[cof(G)]^{\mathrm{T}} = [g^{\alpha\beta}]g \Rightarrow [cof(G)]^{\alpha\beta} = [g^{\alpha\beta}]^{\mathrm{T}}g = g^{\beta\alpha}g$$
$$[cof(G)]^{\alpha\beta} = g^{\beta\alpha}g = g^{\alpha\beta}g. \tag{2.188}$$

Substituting equation (2.188) into equation (2.185,2.1), one can write

$$\partial_\mu g = g g^{\alpha\beta}\partial_\mu g_{\alpha\beta}, \tag{2.189}$$

and using equation (2.174), we find

$$\partial_\mu g = g g^{\alpha\beta}\partial_\mu g_{\alpha\beta} = g g^{\alpha\beta}\left(\Gamma_{\alpha\beta\mu} + \Gamma_{\beta\alpha\mu}\right). \tag{2.190}$$

Substituting the relations

$$\Gamma_{\alpha\beta\mu} = g_{\alpha\nu}\Gamma^{\nu}_{\beta\mu}, \ \Gamma_{\beta\alpha\mu} = g_{\beta\nu}\Gamma^{\nu}_{\alpha\mu}, \tag{2.191}$$

into equation (2.190):

$$\partial_\mu g = g g^{\alpha\beta}\left(g_{\alpha\nu}\Gamma^{\nu}_{\beta\mu} + g_{\beta\nu}\Gamma^{\nu}_{\alpha\mu}\right) = g\left(g^{\alpha\beta}g_{\alpha\nu}\Gamma^{\nu}_{\beta\mu} + g^{\alpha\beta}g_{\beta\nu}\Gamma^{\nu}_{\alpha\mu}\right)$$
$$= g\left(\delta^{\beta}_{\nu}\Gamma^{\nu}_{\beta\mu} + \delta^{\alpha}_{\nu}\Gamma^{\nu}_{\alpha\mu}\right) \Rightarrow \partial_\mu g = g\left(\Gamma^{\beta}_{\beta\mu} + \Gamma^{\alpha}_{\alpha\mu}\right). \tag{2.192}$$

Taking into account that α and β are dummy indices, we can replace β by α in the first term so that

$$\partial_\mu g = 2g\Gamma^{\alpha}_{\alpha\mu} \Rightarrow \Gamma^{\alpha}_{\alpha\mu} = \frac{1}{2g}\partial_\mu g = \frac{1}{2}\partial_\mu \ln |g| = \partial_\mu \ln \sqrt{|g|}. \tag{2.193}$$

Now, for the sake of convenience, if we replace μ by β, we can rewrite the above equation as

$$\Gamma^{\alpha}_{\alpha\beta} = \partial_\beta \ln \sqrt{|g|}. \tag{2.194}$$

The modulus is for the case where the manifold is pseudo-Riemannian, where the metric elements can be negative.

2.12 Local geodesic and Cartesian coordinates

Let us consider a manifold with a coordinates system x^α and a point P on this manifold with coordinates x^{α}_P. We then define a new system of coordinates, x'^{α}, in terms of x^{α}_P and the coordinates x^α as

$$x^{'\alpha} = x^\alpha - x_P^\alpha + \frac{1}{2}\Gamma^\alpha_{\beta\mu}(P)(x^\beta - x_P^\beta)(x^\mu - x_P^\mu), \tag{2.195}$$

where

$$\Gamma^\alpha_{\beta\mu}(P) = \Gamma^\alpha_{\beta\mu}\left(x_p^1, x_p^2, ...x_p^N\right) \tag{2.196}$$

is the affine connections evaluated at point P. We must know how under this transformation the affine connection transformed as we may need to find the derivative of a vector expressed in terms of these new coordinate basis vectors, which depends on the affine connection.

We recall that under coordinate transformation $x^\alpha \to x^{'\alpha}$, the affine connection is transformed according to

$$\Gamma^{'\beta}_{\alpha\mu} = \frac{\partial x^{'\beta}}{\partial x^\nu}\frac{\partial x^\sigma}{\partial x^{'\alpha}}\frac{\partial x^\kappa}{\partial x^{'\mu}}\Gamma^\nu_{\sigma\kappa} + \frac{\partial x^{'\beta}}{\partial x^\nu}\frac{\partial^2 x^\nu}{\partial x^{'\mu}\partial x^{'\alpha}}. \tag{2.197}$$

In order to determine this at point P, we need to differentiate equation (2.195):

$$\begin{aligned}\frac{\partial x^{'\alpha}}{\partial x^\nu} &= \frac{\partial x^\alpha}{\partial x^\nu} + \frac{1}{2}\Gamma^\alpha_{\beta\mu}(P)\frac{\partial}{\partial x^\nu}\{(x^\beta - x_P^\beta)(x^\mu - x_P^\mu)\} \\ &= \frac{\partial x^\alpha}{\partial x^\nu} + \frac{1}{2}\Gamma^\alpha_{\beta\mu}(P)\left\{(x^\mu - x_P^\mu)\frac{\partial x^\beta}{\partial x^\nu} + (x^\beta - x_P^\beta)\frac{\partial x^\mu}{\partial x^\nu}\right\},\end{aligned} \tag{2.198}$$

where we used the fact that $\Gamma^\alpha_{\beta\mu}(P)$, x_P^α, x_P^β, and x_P^μ are constants. Since the coordinates x^α are independent coordinates, we have

$$\frac{\partial x^\alpha}{\partial x^\nu} = \delta_\nu^\alpha, \frac{\partial x^\beta}{\partial x^\nu} = \delta_\nu^\beta, \frac{\partial x^\mu}{\partial x^\nu} = \delta_\nu^\mu, \tag{2.199}$$

so that

$$\frac{\partial x^{'\alpha}}{\partial x^\nu} = \delta_\nu^\alpha + \frac{1}{2}\Gamma^\alpha_{\beta\mu}(P)\{(x^\mu - x_P^\mu)\delta_\nu^\beta + (x^\beta - x_P^\beta)\delta_\nu^\mu\}, \tag{2.200}$$

which simplifies into

$$\frac{\partial x^{'\alpha}}{\partial x^\nu} = \delta_\nu^\alpha + \frac{1}{2}\Gamma^\alpha_{\nu\mu}(P)(x^\mu - x_P^\mu) + \Gamma^\alpha_{\beta\nu}(P)(x^\beta - x_P^\beta). \tag{2.201}$$

Since the summation indices are dummy indices, we can replace β by μ in the second term and write

$$\frac{\partial x^{'\alpha}}{\partial x^\nu} = \delta_\nu^\alpha + \frac{1}{2}\left[\Gamma^\alpha_{\nu\mu}(P)(x^\mu - x_P^\mu) + \Gamma^\alpha_{\mu\nu}(P)(x^\mu - x_P^\mu)\right]. \tag{2.202}$$

We recall, for a torsionless manifold,

$$\Gamma^\alpha_{\nu\mu}(P) = \Gamma^\alpha_{\mu\nu}(P), \tag{2.203}$$

and equation (2.202) becomes

$$\frac{\partial x^{'\alpha}}{\partial x^{\nu}} = \delta_{\nu}^{\alpha} + \Gamma_{\nu\mu}^{\alpha}(P)(x^{\mu} - x_P^{\mu}). \tag{2.204}$$

We are interested in the affine connection at point P. Thus when we are evaluating equation (2.204) at $x^{\mu} = x_P^{\mu}$, we find

$$\left.\frac{\partial x^{'\alpha}}{\partial x^{\nu}}\right|_{P} = \delta_{\nu}^{\alpha}. \tag{2.205}$$

Similarly, one can also show that

$$x^{'\alpha} = x^{\alpha} - x_P^{\alpha} + \frac{1}{2}\Gamma_{\beta\mu}^{\alpha}(P)(x^{\beta} - x_P^{\beta})(x^{\mu} - x_P^{\mu})$$

$$\Rightarrow \frac{\partial x^{'\alpha}}{\partial x^{'\nu}} = \frac{\partial x^{\alpha}}{\partial x^{'\nu}} + \frac{1}{2}\Gamma_{\beta\mu}^{\alpha}(P)\frac{\partial}{\partial x^{'\nu}}[(x^{\beta} - x_P^{\beta})(x^{\mu} - x_P^{\mu})] \tag{2.206}$$

$$\Rightarrow \delta_{\nu}^{\alpha} = \frac{\partial x^{\alpha}}{\partial x^{'\nu}} + \frac{1}{2}\Gamma_{\beta\mu}^{\alpha}(P)\left[(x^{\beta} - x_P^{\beta})\frac{\partial x^{\mu}}{\partial x^{'\nu}} + (x^{\mu} - x_P^{\mu})\frac{\partial x^{\beta}}{\partial x^{'\nu}}\right].$$

The inverse is also given by

$$\left.\frac{\partial x^{\alpha}}{\partial x^{'\nu}}\right|_{P} = \delta_{\nu}^{\alpha}. \tag{2.207}$$

Differentiating equation (2.204) with respect to x^{κ}, we have

$$\frac{\partial x^{'\alpha}}{\partial x^{\kappa}\partial x^{\nu}} = \frac{\partial}{\partial x^{\kappa}}\delta_{\nu}^{\alpha} + \Gamma_{\nu\mu}^{\alpha}(P)\frac{\partial}{\partial x^{\kappa}}(x^{\mu} - x_P^{\mu}) = \frac{\partial}{\partial x^{\kappa}}\delta_{\nu}^{\alpha} + \Gamma_{\nu\mu}^{\alpha}(P)\frac{\partial x^{\mu}}{\partial x^{\kappa}}$$

$$\Rightarrow \frac{\partial x^{'\alpha}}{\partial x^{\kappa}\partial x^{\nu}} = \Gamma_{\nu\mu}^{\alpha}(P)\delta_{\kappa}^{\mu} = \Gamma_{\nu\kappa}^{\alpha}(P). \tag{2.208}$$

Also using equation (2.206), one can write

$$\frac{\partial \delta_{\nu}^{\alpha}}{\partial x^{'\kappa}} = \frac{\partial^2 x^{\alpha}}{\partial x^{'\kappa}\partial x^{'\nu}} + \frac{1}{2}\Gamma_{\beta\mu}^{\alpha}(P)\frac{\partial}{\partial x^{'\kappa}}\left[(x^{\beta} - x_P^{\beta})\frac{\partial x^{\mu}}{\partial x^{'\nu}} + (x^{\mu} - x_P^{\mu})\frac{\partial x^{\beta}}{\partial x^{'\nu}}\right]$$

$$\Rightarrow 0 = \frac{\partial^2 x^{\alpha}}{\partial x^{'\kappa}\partial x^{'\nu}} + \frac{1}{2}\Gamma_{\beta\mu}^{\alpha}(P)\left[\frac{\partial x^{\mu}}{\partial x^{'\nu}}\frac{\partial x^{\beta}}{\partial x^{'\kappa}} + (x^{\beta} - x_P^{\beta})\frac{\partial^2 x^{\mu}}{\partial x^{'\kappa}\partial x^{'\nu}}\right. \tag{2.209}$$

$$\left. + \frac{\partial x^{\mu}}{\partial x^{'\kappa}}\frac{\partial x^{\beta}}{\partial x^{'\nu}} + (x^{\mu} - x_P^{\mu})\frac{\partial^2 x^{\beta}}{\partial x^{'\kappa}\partial x^{'\nu}}\right]$$

so that when this is evaluated at point P, we find

$$\frac{\partial^2 x^{\alpha}}{\partial x^{'\kappa}\partial x^{'\nu}} = -\frac{1}{2}\Gamma_{\beta\mu}^{\alpha}(P)\left[\frac{\partial x^{\mu}}{\partial x^{'\nu}}\frac{\partial x^{\beta}}{\partial x^{'\kappa}} + \frac{\partial x^{\mu}}{\partial x^{'\kappa}}\frac{\partial x^{\beta}}{\partial x^{'\nu}}\right], \tag{2.210}$$

and applying the result in equation (2.207), equation (2.210) leads to

$$\frac{\partial^2 x^\alpha}{\partial x'^\kappa \partial x'^\nu} = -\frac{1}{2}\Big[\Gamma^\alpha_{\beta\mu}(P)\delta^\mu_\nu\delta^\beta_\kappa + \Gamma^\alpha_{\beta\mu}(P)\delta^\mu_\kappa\delta^\beta_\nu\Big] = -\frac{1}{2}[\Gamma^\alpha_{\kappa\nu}(P) + \Gamma^\alpha_{\nu\kappa}(P)]$$

$$\Rightarrow \frac{\partial^2 x^\alpha}{\partial x'^\kappa \partial x'^\nu} = -\Gamma^\alpha_{\kappa\nu}(P) \Rightarrow \frac{\partial^2 x^\nu}{\partial x'^\mu \partial x'^\alpha} = -\Gamma^\nu_{\mu\alpha}(P),$$

(2.211)

where we switched the places for the ν and α indices and replaced κ by μ. Using the results in equations (2.205)–(2.211), the transformation equation for the affine connection,

$$\Gamma'^\beta_{\alpha\mu} = \frac{\partial x'^\beta}{\partial x^\nu}\frac{\partial x^\sigma}{\partial x'^\alpha}\frac{\partial x^\kappa}{\partial x'^\mu}\Gamma^\nu_{\sigma\kappa} + \frac{\partial x'^\beta}{\partial x^\nu}\frac{\partial^2 x^\nu}{\partial x'^\mu \partial x'^\alpha},$$

(2.212)

becomes

$$\Gamma'^\beta_{\alpha\mu}(P) = \delta^\beta_\nu\delta^\sigma_\alpha\delta^\kappa_\mu\Gamma^\nu_{\sigma\kappa}(P) - \delta^\beta_\nu\Gamma^\nu_{\mu\alpha}(P)$$

$$\Rightarrow \Gamma'^\beta_{\alpha\mu}(P) = \Gamma^\beta_{\alpha\mu}(P) - \Gamma^\beta_{\mu\alpha}(P) = 0.$$

(2.213)

The result in equation (2.213) shows that for the coordinate transformation defined by equation (2.195), the affine connection becomes zero at point P. Such coordinates where the affine connection becomes zero at a point P on a manifold is known as *local geodesic coordinates* about P.

In chapter 1 we showed that the conditions for local Cartesian coordinates at a given point P in a pseudo-Riemannian manifold are given by

$$g'_{\alpha\beta}(P) = \eta_{\alpha\beta},$$

(2.214)

$$\frac{\partial g'_{\alpha\beta}(x')}{\partial x'^\mu}\bigg|_P = \partial'_\mu g'_{\alpha\beta}(x')\bigg|_P = 0,$$

(2.215)

where $[\eta_{\alpha\beta}] = diag(\pm 1, \pm 1, \ldots, \pm 1)$. We also know that the number of positive entries (N_+) minus the number of negative entries (N_-) in $[\eta_{\alpha\beta}]$ is the *signature* of the manifold. For geodesic coordinates, equation (2.215) can easily be shown by applying the equation that relates the affine connections with the metric in equation (2.158). For the x'^α coordinates, equation (2.158) can be written as

$$\partial'_\mu g'_{\alpha\beta} = \Gamma'^\nu_{\alpha\mu} g'_{\beta\nu} + \Gamma'^\nu_{\beta\mu} g'_{\alpha\nu}.$$

(2.216)

When this equation is evaluated at point P,

$$\partial'_\mu g'_{\alpha\beta} = \frac{\partial g'_{\alpha\beta}(x')}{\partial x'^\mu}\bigg|_P = \partial'_\mu g'_{\alpha\beta}(x')\bigg|_P = \Gamma'^\nu_{\alpha\mu}(P)g'_{\beta\nu}(P) + \Gamma'^\nu_{\beta\mu}(P)g'_{\alpha\nu}(P),$$

(2.217)

and for a geodesic coordinates the affine connection is zero at point P,

$$\Gamma_{\alpha\mu}^{'\beta}(P) = 0, \tag{2.218}$$

and therefore

$$\left.\frac{\partial g_{\alpha\beta}'(x')}{\partial x^{'\mu}}\right|_P = \partial_\mu' g_{\alpha\beta}'(x')\big|_P = 0. \tag{2.219}$$

It is important to note that for geodesic coordinates the metric does not necessarily satisfy equation (2.214). But we can find coordinates $x^{''\alpha}$ that satisfy equation (2.214) by making a linear transformation for the $x^{'\alpha}$,

$$x^{''\alpha} = X_\beta^\alpha x^{'\beta}, \tag{2.220}$$

where X_β^α are constants.

2.13 The gradient, the divergence, and the curl on a manifold

Before we see how the gradient of a scalar function, the divergence or the curl of a vector function is determined on a manifold, first we need to recall how a vector (like the four-dimensional momentum in Minkowski space-time manifold in equation (2.115)) is partially differentiated with respect to the coordinates. Consider a vector, $\vec{v}$, in terms of its contravariant components:

$$\vec{v} = v^\alpha \hat{e}_\alpha, \tag{2.221}$$

where $\hat{e}_\alpha$ are the coordinate basis vectors. The derivative of this vector with respect to the coordinate x^β can be expressed as

$$\frac{\partial \vec{v}}{\partial x^\beta} = \partial_\beta \vec{v} = \partial_\beta(v^\alpha \hat{e}_\alpha) = \hat{e}_\alpha \partial_\beta(v^\alpha) + v^\alpha \partial_\beta(\hat{e}_\alpha). \tag{2.222}$$

We recall

$$\partial_\beta(\hat{e}_\alpha) = \Gamma_{\alpha\beta}^\mu \hat{e}_\mu, \tag{2.223}$$

so that

$$\partial_\beta \vec{v} = \left(\partial_\beta v^\alpha\right)\hat{e}_\alpha + \Gamma_{\alpha\beta}^\mu v^\alpha \hat{e}_\mu. \tag{2.224}$$

Switching the places for the indices μ and α in the second term, one can write

$$\Gamma_{\alpha\beta}^\mu v^\alpha \hat{e}_\mu = \Gamma_{\mu\beta}^\alpha v^\mu \hat{e}_\alpha = v^\mu \Gamma_{\mu\beta}^\alpha \hat{e}_\alpha, \tag{2.225}$$

so that substituting this into equation (2.224), one finds

$$\partial_\beta \vec{v} = \left(\partial_\beta v^\alpha + \Gamma_{\mu\beta}^\alpha v^\mu\right)\hat{e}_\alpha, \tag{2.226}$$

where the quantity in the bracket,

$$\nabla_\beta v^\alpha = \partial_\beta v^\alpha + \Gamma_{\mu\beta}^\alpha v^\mu, \tag{2.227}$$

is known as the *covariant derivative* of the vector components. Thus the derivative of a vector can be expressed as

$$\partial_\beta \vec{v} = \left(\nabla_\beta v^\alpha\right)\hat{e}_\alpha. \tag{2.228}$$

For geodesic coordinates, where the affine connections vanishes locally, we have

$$\Gamma^\alpha_{\mu\beta} = 0, \tag{2.229}$$

and the covariant derivative in equation (2.227) becomes

$$\nabla_\beta v^\alpha = \partial_\beta v^\alpha, \tag{2.230}$$

which is just the ordinary derivative that we are familiar with.

The covariant derivative of a scalar function: For a scalar function ϕ the covariant derivative is

$$\nabla_\beta \phi = \partial_\beta \phi. \tag{2.231}$$

The gradient: The gradient of a scalar function ϕ is given by

$$\nabla \phi = (\partial_\alpha \phi)\hat{e}^\alpha. \tag{2.232}$$

The divergence: The divergence of a vector function expressed in terms of its contravariant components

$$\vec{v} = v^\alpha \hat{e}_\alpha, \tag{2.233}$$

is given by

$$\nabla \cdot \vec{v} = \nabla_\alpha v^\alpha. \tag{2.234}$$

Using the relation for the covariant derivative of a vector,

$$\nabla_\beta v^\alpha = \partial_\beta v^\alpha + \Gamma^\alpha_{\mu\beta} v^\mu, \tag{2.235}$$

for $\beta = \alpha$, we have

$$\nabla_\alpha v^\alpha = \partial_\alpha v^\alpha + \Gamma^\alpha_{\mu\alpha} v^\mu, \tag{2.236}$$

so that replacing μ by β, one can write

$$\nabla \cdot \vec{v} = \nabla_\alpha v^\alpha = \partial_\alpha v^\alpha + \Gamma^\alpha_{\beta\alpha} v^\beta. \tag{2.237}$$

Using the relation for the affine connection in equation (2.194),

$$\Gamma^\alpha_{\beta\alpha} = \Gamma^\alpha_{\alpha\beta} = \partial_\beta \ln \sqrt{|g|} = \frac{1}{\sqrt{|g|}}\partial_\alpha \sqrt{|g|}, \tag{2.238}$$

we can write

$$\nabla \cdot \vec{v} = \partial_\alpha v^\alpha + v^\alpha \frac{1}{\sqrt{|g|}}\partial_\alpha \sqrt{|g|}, \tag{2.239}$$

which can be put in the form

$$\nabla \cdot \vec{v} = \frac{1}{\sqrt{|g|}} \partial_\alpha \left[v^\alpha \sqrt{|g|} \right]. \tag{2.240}$$

Equation (2.240) shows that the divergence is related to the determinant of the metric tensor.

The Laplacian: We recall that in Euclidean space the Laplacian of the scalar function ϕ is given by

$$\nabla^2 \phi = \nabla \cdot (\nabla \phi). \tag{2.241}$$

Applying equation (2.232), this can be rewritten as

$$\nabla^2 \phi = \nabla \cdot [(\partial_\alpha \phi) \hat{e}^\alpha] = \nabla \cdot \vec{v}, \tag{2.242}$$

where

$$\vec{v} = (\partial_\alpha \phi) \hat{e}^\alpha = v_\alpha \hat{e}^\alpha, \tag{2.243}$$

with $v_\alpha = \partial_\alpha \phi$. In order to apply the relation for the divergence in equation (2.240), we need to express the vector $\vec{v}$ in terms of its contravariant components:

$$\vec{v} = v^\alpha \hat{e}_\alpha. \tag{2.244}$$

We have seen that the index can be raised or lowered using the metric tensor. In this case we want to raise it. Using the contravariant component for the metric tensor, one can write

$$g^{\alpha\beta} v_\beta = v^\alpha, \tag{2.245}$$

so that the vector $\vec{v}$ can be written as

$$\vec{v} = v^\alpha \hat{e}_\alpha = g^{\alpha\beta} v_\beta \hat{e}_\alpha = g^{\alpha\beta} (\partial_\beta \phi) \hat{e}_\alpha. \tag{2.246}$$

Then the Laplacian in equation (2.242) becomes

$$\nabla^2 \phi = \nabla \cdot \vec{v} = \nabla \cdot \left[g^{\alpha\beta} (\partial_\beta \phi) \hat{e}_\alpha \right]. \tag{2.247}$$

Now, applying the relation

$$\nabla \cdot \vec{v} = \frac{1}{\sqrt{|g|}} \partial_\alpha \left[v^\alpha \sqrt{|g|} \right], \tag{2.248}$$

one can write the Laplacian as

$$\nabla^2 \phi = \nabla_\alpha \nabla^\alpha \phi = \frac{1}{\sqrt{|g|}} \partial_\alpha \left[g^{\alpha\beta} (\partial_\beta \phi) \sqrt{|g|} \right]. \tag{2.249}$$

The Laplacian symbol ∇^2 is used in the usual three-dimensional Euclidean space or in any N-dimensional manifold. On the other hand, in a four-dimensional

Minkowski space-time manifold, in relativistic electrodynamics, ∇^2 is replaced by $\square^2$, known as the *d'Alembert operator*.

Curl: We recall for vectors expressed in terms of Cartesian coordinates,

$$\vec{v} = v_x\hat{x} + v_y\hat{y} + v_z\hat{z}, \qquad (2.250)$$

the curl is given by

$$(\nabla \times \vec{v})_x = (\nabla \times \vec{v})_{yz} = \frac{\partial v_z}{\partial y} - \frac{\partial v_y}{\partial z},$$

$$(\nabla \times \vec{v})_y = (\nabla \times \vec{v})_{zx} = \frac{\partial v_x}{\partial z} - \frac{\partial v_z}{\partial x}, \qquad (2.251)$$

$$(\nabla \times \vec{v})_z = (\nabla \times \vec{v})_{xy} = \frac{\partial v_y}{\partial x} - \frac{\partial v_z}{\partial y},$$

which can be rewritten as

$$(\nabla \times \vec{v})_{\alpha\beta} = \frac{\partial v_\beta}{\partial x^\alpha} - \frac{\partial v_\alpha}{\partial x^\beta} = \partial_\alpha v_\beta - \partial_\beta v_\alpha, \qquad (2.252)$$

for $\alpha, \beta = 1, 2, 3$ and $\alpha \neq \beta$, with $x = x^1, y = x^2, z = x^3$ and $v_x = v_1, v_y = v_2, v_z = v_3$. The curl of a vector on a manifold is defined as a rank-2 antisymmetric tensor with components

$$(\text{curl}\,\vec{v})_{\alpha\beta} = \nabla_\alpha v_\beta - \nabla_\beta v_\alpha, \qquad (2.253)$$

where

$$\nabla_\alpha v_\beta = \partial_\alpha v_\beta - \Gamma^\mu_{\beta\alpha} v_\mu, \; \nabla_\beta v_\alpha = \partial_\beta v_\alpha - \Gamma^\mu_{\alpha\beta} v_\mu. \qquad (2.254)$$

Upon substituting equation (2.254) into equation (2.253), one finds

$$(\text{curl}\,\vec{v})_{\alpha\beta} = \partial_\alpha v_\beta - \Gamma^\mu_{\beta\alpha} v_\mu - \partial_\beta v_\alpha + \Gamma^\mu_{\alpha\beta} v_\mu = \partial_\alpha v_\beta - \partial_\beta v_\alpha, \qquad (2.255)$$

for a torsionless manifold $\Gamma^\mu_{\beta\alpha} = \Gamma^\mu_{\alpha\beta}$. Note that equation (2.255) is the same as the expression we had for Cartesian coordinates in a three-dimensional Euclidean space in equation (2.252).

2.14 Intrinsic derivative of a vector along a curve

We may encounter vector fields that depend on a curve instead of the entire or some region of the manifold. In such cases, the curve may be defined by the coordinates x^α that depend on some parameter u (i.e., $x(u)$). Let us consider a vector $\vec{v}$ expressed in terms of its contravariant components. Since the coordinates on this curve depends on the parameter u, the contravariant components for the vector $\vec{v}$ ($v^\alpha(u)$) and the coordinate basis vectors ($\hat{e}_\alpha(x^\alpha(u))$) are also functions of the parameter u:

$$\vec{v}(u) = v^\alpha(u)\hat{e}_\alpha(x). \qquad (2.256)$$

The derivative of this vector along the curve is given by

$$\frac{d}{du}\vec{v}(u) = \frac{d}{du}[v^\alpha(u)\hat{e}_\alpha(x)] = v^\alpha(u)\frac{d\hat{e}_\alpha(u)}{du} + \hat{e}_\alpha(u)\frac{dv^\alpha(u)}{du}. \tag{2.257}$$

Using the relation

$$\frac{d\hat{e}_\alpha}{dx^\beta} = \partial_\beta \hat{e}_\alpha = \Gamma^\kappa_{\alpha\beta}\hat{e}_\kappa, \tag{2.258}$$

we have

$$\frac{d\hat{e}_\alpha(u)}{du} = \left(\partial_\beta\hat{e}_\alpha(u)\right)\frac{dx^\beta}{du} = \frac{dx^\beta}{du}\Gamma^\kappa_{\alpha\beta}\hat{e}_\kappa, \tag{2.259}$$

so that equation (2.257) becomes

$$\frac{d}{du}\vec{v}(u) = v^\alpha\frac{dx^\beta}{du}\Gamma^\kappa_{\alpha\beta}\hat{e}_\kappa + \hat{e}_\alpha\frac{dv^\alpha}{du} = \left(v^\alpha\frac{dx^\beta}{du}\Gamma^\kappa_{\alpha\beta} + \frac{dv^\kappa}{du}\right)\hat{e}_\kappa, \tag{2.260}$$

where we replace the dummy index α by κ in the second term. This can be rewritten as

$$\frac{d}{du}\vec{v}(u) = \left(\frac{dv^\kappa}{du} + \Gamma^\kappa_{\alpha\beta}v^\alpha\frac{dx^\beta}{du}\right)\hat{e}_\kappa, \tag{2.261}$$

which we can put in the form

$$\frac{d}{du}\vec{v}(u) = \frac{Dv^\kappa}{Du}\hat{e}_\kappa, \tag{2.262}$$

where

$$\frac{Dv^\kappa}{Du} = \frac{dv^\kappa}{du} + \Gamma^\kappa_{\alpha\beta}v^\alpha\frac{dx^\beta}{du}, \tag{2.263}$$

which is the *intrinsic (or absolute) derivative* of the contravariant component v^α. Substituting

$$\frac{dv^\kappa}{du} = \left(\partial_\beta v^\kappa\right)\frac{dx^\beta}{du} \tag{2.264}$$

into equation (2.263), we find

$$\frac{Dv^\kappa}{Du} = \frac{dv^\kappa}{du} + \Gamma^\kappa_{\alpha\beta}v^\alpha\frac{dx^\beta}{du} = \left(\partial_\beta v^\kappa + \Gamma^\kappa_{\alpha\beta}v^\alpha\right)\frac{dx^\beta}{du}$$
$$\Rightarrow \frac{Dv^\kappa}{Du} = (\nabla_\beta v^\kappa)\frac{dx^\beta}{du}, \tag{2.265}$$

where we used the relation in equation (2.227). Equation (2.265) gives the intrinsic derivative of the contravariant component of a vector in terms of the derivative of the coordinate x^β with respect to the parameter u.

2.15 Null curves, non-null curves, and affine parameter

We recall that the tangent vector $\vec{t}$ at point p on a manifold is the vector that lies in the tangent space T_p at that point (see figure 2.14), and is given by

$$\vec{t} = \lim_{\delta u \to 0} \frac{\delta \vec{s}}{\delta u} = \frac{d\vec{s}}{du}, \tag{2.266}$$

where $\delta \vec{s}$ is the infinitesimal separation vector between the point P at $x^\alpha(u)$ and another nearby point Q on the curve at $x^\alpha(u + \delta u)$. In a given coordinate system x^α with basis vectors $\hat{e}_\alpha$, we can express the infinitesimal separation vector $d\vec{s}$ as

$$d\vec{s} = dx^\alpha \hat{e}_\alpha, \tag{2.267}$$

so that the tangent vector becomes

$$\vec{t} = \frac{dx^\alpha}{du} \hat{e}_\alpha = t^\alpha \hat{e}_\alpha. \tag{2.268}$$

Recalling that in a pseudo-Riemannian manifold the magnitude of a vector $\vec{v}$ is given by

$$|\vec{v}| = \sqrt{\left| g_{\alpha\beta} v^\alpha v^\beta \right|} = \sqrt{\left| g^{\alpha\beta} v_\alpha v_\beta \right|} = \sqrt{\left| v^\beta v_\beta \right|}, \tag{2.269}$$

in view of this general relation, along with equation (2.268), the magnitude of the tangent vector can be written as

$$|\vec{t}| = \sqrt{\left| g_{\alpha\beta} t^\alpha t^\beta \right|} = \sqrt{\left| g_{\alpha\beta} \frac{dx^\alpha}{du} \frac{dx^\beta}{du} \right|} = \frac{\sqrt{\left| g_{\alpha\beta} dx^\alpha dx^\beta \right|}}{du}. \tag{2.270}$$

Substituting the interval (the distance squared along the curve on the manifold between the two points P and Q)

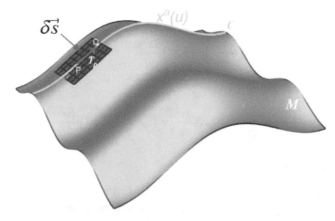

Figure 2.14. A tangent vector at point P in tangent space.

$$(ds)^2 = g_{\alpha\beta}dx^\alpha dx^\beta, \tag{2.271}$$

into equation (2.270), one finds

$$\left| \vec{t} \right| = \left| \frac{ds}{du} \right| = \begin{cases} 0, & \text{null vector,} \\ \neq 0, & \text{non - null vector.} \end{cases} \tag{2.272}$$

For a non-null tangent vector, the infinitesimal length ds at all points on the curve must be different from zero and therefore depends on the parameter u at all points on the curve $s = s(u)$. If the parameter u and the length s are related by the linear equation

$$u = ms + k \tag{2.273}$$

for $m, k \neq 0$, the parameter u is called the *affine parameter* on the curve. However, when the tangent vector is a null vector,

$$| t | = \left| \frac{ds}{du} \right| = 0, \tag{2.274}$$

at all points on the curve and the distance s does not depend on the parameter u, clearly we cannot use it as an *affine parameter* since it does not satisfy the condition in equation (2.273). In such cases, we need to find a parameter known as the *privileged family of affine parameters* (figure 2.14).

Example 2.6. Let us reconsider 'the three guys' scenario: the man on Earth (S frame) and the two aliens (Alien 1 and Alien 2). In the S' frame (Alien 2) that is moving with a speed v along the positive x-axis (measured by the man) as shown in figure 2.15. Suddenly, an event occurs: the appearance of Alien 1. This event is recorded by the two observers in the S frame (the man) and S' frame (Alien 2). The man records this event in the Minkowski space-time at the point described by the coordinates

$$(t, x, y, z) \rightarrow (ct, x, y, z) \rightarrow (x^0, x^1, x^2, x^3), \tag{2.275}$$

and Alien 2 records this same event at

$$(t', x', y', z') \rightarrow (ct', x', y', z') \rightarrow (x'^0, x'^1, x'^2, x'^3). \tag{2.276}$$

These two coordinates are related by the Lorentz transformation for $x'^\alpha \rightarrow x^\alpha$,

$$x^0 = \gamma_v x'^0 + \gamma_v \beta x'^1, \ x^1 = \gamma_v \beta x'^0 + \gamma_v x'^1, \ x^2 = x'^2, \ x^3 = x'^3,$$

$$\Rightarrow [\Lambda^\alpha_\kappa] = \left[\frac{\partial x^\alpha}{\partial x'^\kappa} \right] = \begin{bmatrix} \gamma_v & \gamma_v\beta & 0 & 0 \\ \gamma_v\beta & \gamma_v & 0 & 0 \\ 0 & 0 & 1 & 0 \\ 0 & 0 & 0 & 1 \end{bmatrix}, \tag{2.277}$$

and for $x^\alpha \rightarrow x'^\alpha$,

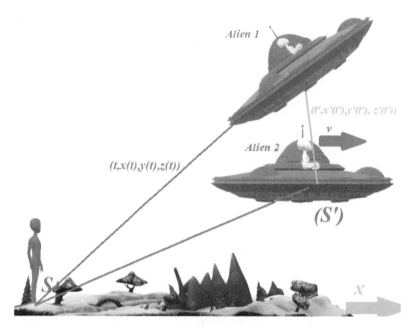

Figure 2.15. An event (appearance of Alien 1) recorded by two observers, an observer in the S frame (a man on Earth) and an observer in the S' frame (Alien 2 moving with a velocity v along the positive x-axis as measured by an observer in the S frame).

$$x'^0 = \gamma_v x^0 - \gamma_v \beta x^1, \; x'^1 = -\gamma_v \beta x^0 + \gamma_v x^1, \; x'^2 = x^2, \; x'^3 = x^3,$$

$$\Rightarrow [\Lambda^{'\alpha}_{\kappa}] = \left[\frac{\partial x'^\alpha}{\partial x^\kappa} \right] = \begin{bmatrix} \gamma_v & -\gamma_v \beta & 0 & 0 \\ -\gamma_v \beta & \gamma_v & 0 & 0 \\ 0 & 0 & 1 & 0 \\ 0 & 0 & 0 & 1 \end{bmatrix}, \tag{2.278}$$

where

$$\beta = \frac{v}{c}, \; \gamma_v = \frac{1}{\sqrt{1 - \dfrac{v^2}{c^2}}}. \tag{2.279}$$

Suppose Alien 1 has been accelerating and the man in the S frame measures the velocity (three-velocity) of Alien 1, at time t (measured by his wrist watch), to be

$$\vec{v}_1(t) = \frac{dx}{dt}\hat{x} + \frac{dy}{dt}\hat{y} + \frac{dz}{dt}\hat{z} = (v_{1x}, v_{1y}, v_{1z}) = (v^1, v^2, v^3), \tag{2.280}$$

along the direction shown in figure 2.15. At the same instant when the wrist watch of Alien 2 ticks a time t', the velocity of Alien 1 is

$$\vec{v}_1'(t') = \frac{dx'}{dt'}\hat{x}' + \frac{dy'}{dt'}\hat{y}' + \frac{dz'}{dt'}\hat{z}' = \left(v_{1x}', v_{1y}', v_{1z}'\right) = (v'^1, v'^2, v'^3). \tag{2.281}$$

In the Minkowski space-time the trajectory of Alien 1 forms the worldline (a curve on the manifold). This curve can be defined in terms of some parameter. For a massive particle this parameter can be the proper time τ, which is defined by

$$\left(\frac{ds}{d\tau}\right)^2 = c^2, \tag{2.282}$$

where ds is an infinitesimal displacement on the worldline. From this equation, one can easily see that since $\tau = cs + k$, τ is the affine parameter. The worldline for Alien 1 can then be parametrized by the proper time and defined by the coordinates $x^\alpha(\tau)$. Then the four-velocity $\vec{V}_1(\tau)$ of Alien 1 is the tangent vector on the worldline determined by the covariant derivative

$$\vec{V}_1(\tau) = \left(\frac{dx^\alpha}{d\tau} + \Gamma^\alpha_{\mu\beta}x^\mu\frac{dx^\beta}{d\tau}\right)\hat{e}_\alpha, \tag{2.283}$$

as measured by the observer (the man) in the S frame. Similarly, for an observer in the S' frame (Alien 2) the four-velocity $\vec{V}_1'(\tau)$ is given by

$$\vec{V}_1'(\tau) = \left(\frac{dx'^\alpha}{d\tau} + \Gamma'^\alpha_{\mu\beta}x'^\mu\frac{dx'^\beta}{d\tau}\right)\hat{e}_\alpha'. \tag{2.284}$$

An infinitesimal time measured by a clock in the S frame (dt) (i.e., the man's wrist watch) is related to the proper time ($d\tau$) (Alien 1's wrist watch) by

$$d\tau = \frac{dt}{\gamma_{v_1}}, \tag{2.285}$$

where

$$\gamma_{v_1} = \frac{1}{\sqrt{1 - \dfrac{v_1^2}{c^2}}}. \tag{2.286}$$

Note that v_1 is the magnitude of the velocity ('three-velocity') of Alien 1 in equation (2.280) as measured by an observer in the S frame. Similarly, one can also write for the time dt' in the S' frame (by Alien 2's wrist watch)

$$d\tau = \frac{dt'}{\gamma_{v_1'}}, \tag{2.287}$$

where

$$\gamma_{v_1'} = \frac{1}{\sqrt{1 - \dfrac{v_1'^2}{c^2}}}, \tag{2.288}$$

and v_1' is the magnitude of the three-velocity of Alien 1 in equation (2.281) as measured by an observer in the S' frame (Alien 2).

(a) Using

$$\Gamma_{\alpha\beta\mu} = \frac{1}{2}\left(\partial_\beta g_{\mu\alpha} + \partial_\mu g_{\alpha\beta} - \partial_\alpha g_{\beta\mu}\right), \tag{2.289}$$

show that the affine connection in both x^α and x'^α coordinates is zero,

$$\Gamma^\alpha_{\mu\beta} = \Gamma'^\alpha_{\mu\beta} = 0, \tag{2.290}$$

and the four-velocity of Alien 1 as measured by the man in the S frame and Alien 2 in the S' frame are given by

$$\vec{V_1}(\tau) = \frac{dx^\alpha}{d\tau}\hat{e}_\alpha \text{ and } \vec{V_1'}(\tau) = \frac{dx'^\alpha}{d\tau}\hat{e}_\alpha', \tag{2.291}$$

respectively.

(b) Find the four-velocity of Alien 1 as observed by the man in the S frame, $\vec{V_1}$.
(c) Find the four-velocity of Alien 1 as observed by Alien 2 in the S' frame, $\vec{V_1'}$.
(d) Show that the three-velocity measured by the S frame and S' frame are related by

$$v_{1x}' = \frac{v_{1x} - v}{1 - \frac{v}{c^2}v_{1x}}, \ v_{1y}' = \frac{v_{1y}\sqrt{1 - \frac{v^2}{c^2}}}{1 - \frac{v}{c^2}v_{1x}}, \ v_{1z}' = \frac{v_{1z}\sqrt{1 - \frac{v^2}{c^2}}}{1 - \frac{v}{c^2}v_{1z}}. \tag{2.292}$$

Solution:

(a) We recall that for Minkowski space-time the covariant component of the metric tensor in the S and S' frames (see example 2.5) is

$$\left[g_{\alpha\beta}'\right] = \begin{bmatrix} 1 & 0 & 0 & 0 \\ 0 & -1 & 0 & 0 \\ 0 & 0 & -1 & 0 \\ 0 & 0 & 0 & -1 \end{bmatrix} = [g_{\alpha\beta}] \tag{2.293}$$

We can easily see that

$$\Gamma^\alpha_{\mu\beta} = g^{\alpha\kappa}\Gamma_{\kappa\beta\mu} = \frac{1}{2}g^{\alpha\kappa}\left(\partial_\beta g_{\mu\kappa} + \partial_\mu g_{\kappa\beta} - \partial_\kappa g_{\beta\mu}\right) = 0, \tag{2.294}$$

$$\Gamma'^\alpha_{\mu\beta} = g'^{\alpha\kappa}\Gamma'_{\kappa\beta\mu} = \frac{1}{2}g'^{\alpha\kappa}\left(\partial_\beta' g_{\mu\kappa}' + \partial_\mu' g_{\kappa\beta}' - \partial_\kappa' g_{\beta\mu}'\right) = 0. \tag{2.295}$$

Applying this result, one can then write the four velocities in the S frame and S' frame as

$$\vec{V}_1(\tau) = \left(\frac{dx^\alpha}{d\tau} + \Gamma^\alpha_{\mu\beta} x^\mu \frac{dx^\beta}{d\tau} \right) \hat{e}_\alpha = \frac{dx^\alpha}{d\tau} \hat{e}_\alpha = V^\alpha \hat{e}_\alpha \qquad (2.296)$$

and

$$\vec{V}_1'(\tau) = \left(\frac{dx'^\alpha}{d\tau} + \Gamma'^\alpha_{\mu\beta} x'^\mu \frac{dx'^\beta}{d\tau} \right) \hat{e}_\alpha' = \frac{dx'^\alpha}{d\tau} \hat{e}_\alpha' = V'^\alpha \hat{e}_\alpha', \qquad (2.297)$$

respectively.

(b) Using the result in part (a) the contravariant component of the four-velocity can be written as

$$\vec{V}_1(\tau) = \frac{dx^\alpha}{d\tau} \hat{e}_\alpha = \left[\frac{d(ct)}{d\tau}, \frac{dx^1}{d\tau}, \frac{dx^2}{d\tau}, \frac{dx^3}{d\tau} \right], \qquad (2.298)$$

so that upon substituting

$$d\tau = \frac{dt}{\gamma_{v_1}},$$

one finds

$$[V^\alpha] = \left[\frac{dx^\alpha}{d\tau} \right] = \gamma_{v_1} \left[c, \frac{dx^1}{dt}, \frac{dx^2}{dt}, \frac{dx^3}{dt} \right] = \gamma_{v_1}[c, \vec{v}_1]$$

$$\Rightarrow \begin{bmatrix} V^0 \\ V^1 \\ V^2 \\ V^3 \end{bmatrix} = \gamma_{v_1} \begin{bmatrix} c \\ v_{1x} \\ v_{1y} \\ v_{1z} \end{bmatrix}. \qquad (2.299)$$

(c) In the S' frame the four-velocity is given by

$$\vec{V}_1'(\tau) = \frac{dx'^\alpha}{d\tau} \hat{e}_\alpha' = V'^\alpha \hat{e}_\alpha'. \qquad (2.300)$$

Recalling that

$$d\tau = \frac{dt'}{\gamma_{v_1'}}, \qquad (2.301)$$

one can write

$$\left[\frac{dx'^\alpha}{d\tau} \right] = \gamma_{u'} \left[c, \frac{dx'^1}{dt'}, \frac{dx'^2}{dt'}, \frac{dx'^3}{dt'} \right] = \gamma_{v_1'}[c, \vec{v}_1']$$

$$\Rightarrow \begin{bmatrix} V'^0 \\ V'^1 \\ V'^2 \\ V'^3 \end{bmatrix} = \gamma_{v_1'} \begin{bmatrix} c \\ v_{1x}' \\ v_{1y}' \\ v_{1z}' \end{bmatrix}. \qquad (2.302)$$

(d) The contravariant components for the four-velocity can be determined using the relation

$$V'^{\alpha} = \vec{V}'_1(\tau) \cdot \hat{e}'^{\alpha}. \tag{2.303}$$

Since the four-vector is independent of the coordinate system,

$$\vec{V}'_1(\tau) = V'^{\beta}_1 \hat{e}'_{\beta} = \vec{V}_1(\tau) = V^{\beta}_1 \hat{e}_{\beta}, \tag{2.304}$$

one can write

$$V'^{\alpha} = V^{\beta}_1 \hat{e}_{\beta} \cdot \hat{e}'^{\alpha}, \tag{2.305}$$

and using the transformation

$$\hat{e}'^{\alpha} = \frac{\partial x'^{\alpha}}{\partial x^{\eta}} \hat{e}^{\eta} = \Lambda'^{\alpha}_{\eta} \hat{e}^{\eta}, \tag{2.306}$$

we find

$$V'^{\alpha} = V^{\beta}_1 \hat{e}_{\beta} \cdot \Lambda'^{\alpha}_{\eta} \hat{e}^{\eta} = \Lambda'^{\alpha}_{\eta} V^{\beta}_1 \delta^{\eta}_{\beta} = \Lambda'^{\alpha}_{\beta} V^{\beta}_1, \tag{2.307}$$

or in a matrix form:

$$[V'^{\alpha}] = \left[\Lambda'^{\alpha}_{\beta}\right][V^{\beta}_1]. \tag{2.308}$$

Applying the result from problem 3 in chapter 1:

$$[\Lambda'^{\alpha}_{\kappa}] = \left[\frac{\partial x'^{\alpha}}{\partial x^{\kappa}}\right] = \begin{bmatrix} \gamma_v & -\gamma_v\beta & 0 & 0 \\ -\gamma_v\beta & \gamma_v & 0 & 0 \\ 0 & 0 & 1 & 0 \\ 0 & 0 & 0 & 1 \end{bmatrix}, \tag{2.309}$$

we find

$$\begin{bmatrix} V'^0 \\ V'^1 \\ V'^2 \\ V'^3 \end{bmatrix} = \begin{bmatrix} \gamma_v & -\gamma_v\beta & 0 & 0 \\ -\gamma_v\beta & \gamma_v & 0 & 0 \\ 0 & 0 & 1 & 0 \\ 0 & 0 & 0 & 1 \end{bmatrix} \begin{bmatrix} V^0 \\ V^1 \\ V^2 \\ V^3 \end{bmatrix}. \tag{2.310}$$

Upon substituting equations (2.299) and (2.302), from the results in parts (b) and (c), into equation (2.310), one can write

$$\gamma_{v_1'}\begin{bmatrix} c \\ v_{1x}' \\ v_{1y}' \\ v_{1z}' \end{bmatrix} = \begin{bmatrix} \gamma_{v_1}\gamma_v & -\gamma_{v_1}\gamma_v\beta & 0 & 0 \\ -\gamma_{v_1}\gamma_v\beta & \gamma_{v_1}\gamma_v & 0 & 0 \\ 0 & 0 & \gamma_{v_1} & 0 \\ 0 & 0 & 0 & \gamma_{v_1} \end{bmatrix}\begin{bmatrix} c \\ v_{1x} \\ v_{1y} \\ v_{1z} \end{bmatrix}$$

$$= \begin{bmatrix} \gamma_{v_1}\gamma_v(c - v_{1x}\beta) \\ \gamma_{v_1}\gamma_v(v_{1x} - c\beta) \\ \gamma_{v_1}v_{1y} \\ \gamma_{v_1}v_{1z} \end{bmatrix},$$

$$(2.311)$$

from which follows:

$$\begin{bmatrix} \gamma_{v_1'}c \\ \gamma_{v_1'}v_{1x}' \\ \gamma_{v_1'}v_{1y}' \\ \gamma_{v_1'}v_{1z}' \end{bmatrix} = \begin{bmatrix} \gamma_{v_1}\gamma_v(c - v_{1x}\beta) \\ \gamma_{v_1}\gamma_v(v_{1x} - c\beta) \\ \gamma_{v_1}v_{1y} \\ \gamma_{v_1}v_{1z} \end{bmatrix}. \tag{2.312}$$

From this equation, one can see that

$$\gamma_{v_1'} = \gamma_{v_1}\gamma_v\left(1 - \frac{v_{1x}\beta}{c}\right), \quad v_{1x}' = \frac{\gamma_{v_1}\gamma_v}{\gamma_{v_1'}}(v_{1x} - c\beta),$$

$$v_{1y}' = \frac{\gamma_{v_1}}{\gamma_{v_1'}}v_{1y}, \quad v_{1z}' = \frac{\gamma_{v_1}}{\gamma_{v_1'}}v_{1z}. \tag{2.313}$$

Substituting

$$\gamma_{v_1'} = \gamma_{v_1}\gamma_v\left(1 - \frac{v_{1x}\beta}{c}\right), \tag{2.314}$$

in the expressions for v_{1x}', v_{1y}', and v_{1z}',

$$v_{1x}' = \frac{v_{1x} - c\beta}{1 - \dfrac{v_{1x}\beta}{c}}, \quad v_{1y}' = \frac{1}{\gamma_v}\frac{v_{1y}}{1 - \dfrac{v_{1x}\beta}{c}}, \quad v_{1z}' = \frac{1}{\gamma_v}\frac{v_{1z}}{1 - \dfrac{v_{1x}\beta}{c}}.$$

Now, substituting

$$\beta = \frac{v}{c}, \quad \gamma = \frac{1}{\sqrt{1 - \dfrac{v^2}{c^2}}} \tag{2.315}$$

in equation (2.313), one finds

$$v'_{1x} = \frac{v_{1x} - v}{1 - \frac{v}{c^2}v_{1x}}, \; u'_y = \frac{v_{1y}\sqrt{1 - \frac{v^2}{c^2}}}{1 - \frac{v}{c^2}v_{1x}}, \; u'_z = \frac{v_{1z}\sqrt{1 - \frac{v^2}{c^2}}}{1 - \frac{v}{c^2}v_{1x}}. \tag{2.316}$$

Example 2.7. The four-momentum for a massive particle.

In the previous example we saw that a particle's velocity in the Minkowski space-time is described by the four-velocity, $\vec{V}_1$:

$$\vec{V}_1 = v^\alpha \hat{e}_\alpha = \frac{dx^\alpha}{d\tau}\hat{e}_\alpha = \gamma_{v_1}\left[c, \frac{dx^1}{d\tau}, \frac{dx^2}{d\tau}, \frac{dx^3}{d\tau}\right] = \gamma_{v_1}[c, \vec{v}_1] \tag{2.317}$$

in the S frame. Now let us consider Alien 1 in the spaceship traveling with a velocity (see figure 2.16)

$$\vec{v}_1 = \frac{dx}{dt}\hat{x} + \frac{dy}{dt}\hat{y} + \frac{dz}{dt}\hat{z} = (v_{1x}, v_{1y}, v_{1z}) = (v^1, v^2, v^3), \tag{2.318}$$

as measured by an observer in the $\vec{S}$ frame. Let the rest mass of the particle, that I prefer to call the 'proper mass', be m_0. The proper mass is the mass measured by an observer in a reference frame moving with the same velocity as the particle, which is known as the instantaneous inertial frame (IIF). The four-momentum $\vec{P}$ in the S frame is defined in terms of the proper mass (rest mass):

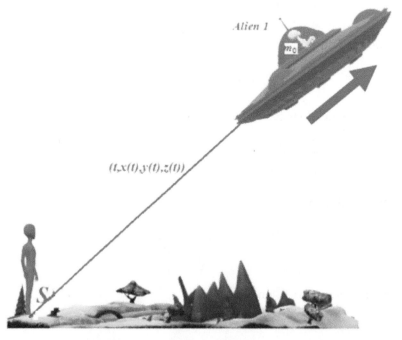

Figure 2.16. A spaceship traveling with a velocity $\vec{v}_1$.

$$\vec{P} = P^\alpha \hat{e}_\alpha = m_0 V^\alpha \hat{e}_\alpha.$$

(a) From this relation show that the relativistic energy is given by

$$E = \frac{E_0}{\sqrt{1 - \dfrac{v_1^2}{c^2}}} \tag{2.319}$$

and momentum

$$\vec{p} = \frac{m_0 \vec{v}_1}{\sqrt{1 - \dfrac{u^2}{c^2}}}, \tag{2.320}$$

where

$$E_0 = m_0 c^2 \tag{2.321}$$

is the rest-mass energy.

(b) The magnitude of the relativistic momentum is given by

$$p = \frac{1}{c}\sqrt{E^2 - E_0^2}. \tag{2.322}$$

Solution:

(a) Applying the result in example 2.6 one can write the four momentum as

$$\vec{P} = P^\alpha \hat{e}_\alpha = m_0 V^\alpha \hat{e}_\alpha = \left[m_0 \frac{dx^\alpha}{d\tau} \right] = \gamma_{v_1}\left[m_0 c, m_0 \frac{dx^1}{dt}, m_0 \frac{dx^2}{dt}, m_0 \frac{dx^3}{dt} \right] \tag{2.323}$$

$$\Rightarrow \vec{P} = \gamma_{v_1}[m_0 c, m_0 \vec{v}_1].$$

Note that here $\vec{u}$ represent the three-velocity. This can be rewritten as

$$[P^\alpha] = [P^0, P^1, P^2, P^3] = \gamma_{v_1}\left[\frac{E_0}{c}, m_0 \vec{v}_1 \right], \tag{2.324}$$

where we introduced the rest-mass energy, which I also prefer to term the 'proper energy', as

$$E_0 = m_0 c^2. \tag{2.325}$$

Recalling that

$$\gamma_{v_1} = \frac{1}{\sqrt{1 - \dfrac{v_1^2}{c^2}}}, \tag{2.326}$$

one can write

$$E = cP^0 = \gamma_{v_1} E_0 = \frac{E_0}{\sqrt{1 - \frac{v_1^2}{c^2}}} = \frac{m_0 c^2}{\sqrt{1 - \frac{v_1^2}{c^2}}}, \tag{2.327}$$

which is *the relativistic energy* and

$$\overrightarrow{p} = [P^1, P^2, P^3] = \gamma_{v_1} m_0 \overrightarrow{v}_1 = \frac{m_0 \overrightarrow{v}_1}{\sqrt{1 - \frac{v_1^2}{c^2}}}, \tag{2.328}$$

is the *the relativistic momentum (three-momentum)*.

$$\vec{p} = [P^1, P^2, P^3] = \frac{m_0 \vec{v}_1}{\sqrt{1 - \frac{v_1^2}{c^2}}} \tag{2.329}$$

is the three-momentum (the relativistic momentum).

(b) The magnitude of the four-momentum can be determined using the inner product:

$$\vec{P} \cdot \vec{P} = P^\alpha P_\alpha = m_0^2 v^\alpha v_\alpha = m_0^2 \left(\frac{ds}{d\tau}\right)^2 = m_0^2 c^2 = \left(\frac{E_0}{c}\right)^2. \tag{2.330}$$

Using the metric tensor for the Minkowski space-time, one can also write

$$P_\alpha = \eta_{\alpha\beta} P^\beta,$$

so that

$$\vec{P} \cdot \vec{P} = P^\alpha P_\alpha = P p^\alpha \eta_{\alpha\beta} P^\beta$$

$$= [P^0 \ P^1 \ P^2 \ P^3] \begin{bmatrix} 1 & 0 & 0 & 0 \\ 0 & -1 & 0 & 0 \\ 0 & 0 & -1 & 0 \\ 0 & 0 & 0 & -1 \end{bmatrix} \begin{bmatrix} P^0 \\ P^1 \\ P^2 \\ P^3 \end{bmatrix}$$

$$= [P^0 \ -P^1 \ -P^2 \ -P^3] \begin{bmatrix} P^0 \\ P^1 \\ P^2 \\ P^3 \end{bmatrix} = (P^0)^2 - (P^1)^2 - (P^2)^2 - (P^3)^2$$

$$\Rightarrow \vec{P} \cdot \vec{P} = \left(\frac{E}{c}\right)^2 - \vec{p} \cdot \vec{p} = \left(\frac{E}{c}\right)^2 - p^2.$$

(2.331)

From these two expressions one can easily find

$$\left(\frac{E}{c}\right)^2 - p^2 = \left(\frac{E_0}{c}\right)^2 \Rightarrow p = \frac{1}{c}\sqrt{E^2 - E_0^2}. \tag{2.332}$$

This is the magnitude of the three-momentum in the special theory of relativity, which you may have seen in introductory physics or modern physics courses.

Example 2.8. Relativistic equation of motion for a massive particle.

The equation of motion for a massive particle is given by Newton's second law in terms of the four-momentum:

$$\vec{F} = \frac{d\vec{P}}{d\tau} = f^\alpha \hat{e}_\alpha = f'^\alpha \hat{e}'_\alpha, \tag{2.333}$$

which gives *the four-force*. Applying the result we determined for the intrinsic derivative of a vector in terms of its contravariant components on a curve, C, parametrized by τ, one can write

$$\frac{d\vec{P}}{d\tau} = \frac{Dp^\kappa}{D\tau} \hat{e}_\kappa, \tag{2.334}$$

where

$$\frac{Dp^\kappa}{D\tau} = \frac{dp^\kappa}{d\tau} + \Gamma^\kappa_{\alpha\beta} p^\alpha \frac{dx^\beta}{d\tau}, \tag{2.335}$$

For the Minkowski space-time, we have shown in example 2.6 that the affine connection $\Gamma^\alpha_{\kappa\beta} = 0$, everywhere. Therefore, the four-momentum becomes

$$\vec{F} = \frac{d\vec{P}}{d\tau} = \frac{dp^\alpha}{d\tau} \hat{e}_\alpha, \tag{2.336}$$

so that the components of the four-force can be expressed as

$$f^\alpha = \vec{F} \cdot \hat{e}^\alpha = \frac{dp^\beta}{d\tau} \hat{e}_\beta \cdot \hat{e}^\alpha = \frac{dp^\beta}{d\tau} \delta^\alpha_\beta = \frac{dp^\alpha}{d\tau}. \tag{2.337}$$

(a) Show that the four-force on a massive particle at a given time t as measured by the man in the S frame can be written as

$$\vec{F} = \gamma_{v_1} \left[\frac{\vec{f} \cdot \vec{v}_1}{c}, \vec{f} \right], \tag{2.338}$$

where $\vec{v}_1$ is the three-velocity of the particle and $\vec{f}$ is the three-force acting on the particle as measured by the man in the S frame at time t.

(b) When the force acting on the particle does not change the mass of the particle, m,

$$\frac{dm}{d\tau} = 0 \Rightarrow m = m_0, \tag{2.339}$$

the force is referred to as a *pure force*. Show that for a pure force

$$\vec{V}_1 \cdot \vec{F} = 0,$$

$$\vec{A} = [a^\alpha] = \gamma_{v_1}\left[0, \gamma_{v_1}\frac{d\vec{v}_1}{dt}\right],$$

(2.340)

where $\vec{V}_1$ is the four-velocity, $\vec{F}$ is the four-force, $\vec{A}$ is the four-acceleration as measured by the man in the S frame at time t.

(c) Suppose the particle is a charged particle with charge q and rest mass m_0 moving in a uniform electric field $\vec{E}_e$ and magnetic field $\vec{B}_m$. This particle would experience a Lorentz force given by

$$\vec{f} = q\vec{E}_e + q(\vec{v} \times \vec{B}_m).$$

(2.341)

Show that the Lorentz force is a pure force.

Solution:

(a) We recall that for a particle traveling with a speed, $\vec{v}_1$, the proper time is given by

$$d\tau = \frac{dt}{\gamma_{v_1}},$$

(2.342)

and we may write the four-force as

$$[f^\alpha] = \left[\frac{dp^\alpha}{d\tau}\right] = \gamma_{v_1}\left[\frac{dp^\alpha}{dt}\right] = \gamma_{v_1}\frac{d}{dt}\left[\frac{E}{c}, \vec{p}\right] = \gamma_{v_1}\left[\frac{1}{c}\frac{dE}{dt}, \frac{d\vec{p}}{dt}\right].$$

(2.343)

We recall *the work energy theorem*,

$$dE = dW = \vec{f} \cdot d\vec{r},$$

(2.344)

where $d\vec{r}$ is the infinitesimal displacement of a massive particle in the S frame. In this frame if the particle has moved this displacement in a time interval, dt, we can write

$$\frac{dE}{dt} = \frac{dW}{dt} = \vec{f} \cdot \frac{d\vec{r}}{dt} = \vec{f} \cdot \vec{v}_1,$$

(2.345)

where $\vec{v}_1$ is the *three-velocity* and

$$\vec{f} = \frac{d\vec{p}}{dt}$$

(2.346)

is the *three-force*. Thus we can write the four-force as

$$\vec{F} = \gamma_{v_1}\left[\frac{\vec{f} \cdot \vec{v}_1}{c}, \vec{f}\right].$$

(2.347)

(b) We recall the energy and the three-momentum of a massive particle moving with a velocity, $\vec{v}_1$,

$$E = \frac{E_0}{\sqrt{1 - \dfrac{v_1^2}{c^2}}} = \frac{m_0 c^2}{\sqrt{1 - \dfrac{v_1^2}{c^2}}}, \quad \vec{p} = \gamma_{v_1} m_0 \vec{v}_1 = \frac{m_0 \vec{v}_1}{\sqrt{1 - \dfrac{v_1^2}{c^2}}}, \tag{2.348}$$

and the four-momentum

$$\vec{P} = p^\kappa \hat{e}_\kappa = \left[\frac{E}{c}, \vec{p} \right]. \tag{2.349}$$

Consider the inner product of the four-velocity and the four-force of a massive particle:

$$\vec{V}_1 \cdot \vec{F} = \vec{V}_1 \cdot \frac{d\vec{P}}{d\tau}. \tag{2.350}$$

If we write the four-momentum in terms of the mass, m, and the four-velocity $\vec{V}_1$ as

$$\vec{P} = m\vec{V}_1 = mv^\alpha \hat{e}_\alpha, \tag{2.351}$$

where m is the relativistic mass, then we have

$$\vec{V}_1 \cdot \vec{F} = \vec{V}_1 \cdot \frac{d\vec{P}}{d\tau} = \vec{V}_1 \cdot \frac{d}{d\tau}\left(m\vec{V}_1\right) = m\vec{V}_1 \cdot \frac{d\vec{V}_1}{d\tau} + \vec{V}_1 \cdot \vec{V}_1 \frac{dm}{d\tau}, \tag{2.352}$$

and noting that for the four-velocity

$$\vec{V}_1 \cdot \vec{V}_1 = c^2 \Rightarrow \frac{d}{d\tau}\left(\vec{V}_1 \cdot \vec{V}_1\right) = 0 \Rightarrow \vec{V}_1 \cdot \frac{d\vec{V}_1}{d\tau} = 0,$$

one can write

$$\vec{V}_1 \cdot \vec{F} = c^2 \frac{dm}{d\tau}. \tag{2.353}$$

Then for a pure force equation (2.353) becomes

$$\vec{V}_1 \cdot \vec{F} = 0, \tag{2.354}$$

and also

$$\begin{aligned} \frac{dE}{d\tau} &= \frac{d}{d\tau}(mc^2) = 0, \\ \frac{d\vec{p}}{d\tau} &= \frac{d(m\vec{v}_1)}{d\tau} = m_0 \frac{d\vec{v}_1}{d\tau} = \gamma_{v_1} \frac{d\vec{p}}{dt}. \end{aligned} \tag{2.355}$$

Then the four-acceleration of a particle acted upon by a pure force becomes

$$[a^\alpha] = \frac{[f^\alpha]}{m_0} = \frac{1}{m_0}\left[\frac{dp^\alpha}{d\tau}\right] = \frac{1}{m_0}\frac{d}{d\tau}\left[\frac{E}{c}, \vec{p}\right]$$

$$= \frac{1}{m_0}\left[\frac{1}{c}\frac{dE}{d\tau}, \frac{d\vec{p}}{d\tau}\right] \Rightarrow [a^\alpha] = \frac{\gamma_{v_1}}{m_0}\left[0, \frac{d\vec{p}}{dt}\right]. \tag{2.356}$$

Using the three-momentum

$$\vec{p} = \frac{m_0\vec{v}_1}{\sqrt{1 - \dfrac{v_1^2}{c^2}}} = \gamma_{v_1}m_0\vec{v}_1,$$

we find

$$[a^\alpha] = \gamma_{v_1}\left[0, \gamma_{v_1}\frac{d\vec{v}_1}{dt}\right]. \tag{2.357}$$

This means the four-acceleration becomes the three-acceleration for a pure force. But, generally, for a force that is not a pure force the acceleration is given by

$$[a^\alpha] = \gamma_u\left[c\frac{d\gamma_u}{dt}, \gamma_u\vec{a} + \vec{u}\frac{d\gamma_u}{dt}\right]. \tag{2.358}$$

(c) Noting that for the Lorentz force

$$\vec{f} = q\vec{E}_e + q\left(\vec{v} \times \vec{B}_m\right), \tag{2.359}$$

we find

$$\vec{v}_1 \cdot \vec{f} = q\left(\vec{v}_1 \cdot \vec{E}_e\right) + q\vec{v}_1 \cdot \left(\vec{v}_1 \times \vec{B}_m\right) = q\left(\vec{v}_1 \cdot \vec{E}_m\right), \tag{2.360}$$

where we used

$$\vec{v}_1 \cdot \left(\vec{v}_1 \times \vec{B}_m\right) = 0.$$

Then for the four-force,

$$\vec{F} = \gamma_{v_1}\left[\frac{\vec{f} \cdot \vec{v}_1}{c}, \vec{f}\right], \tag{2.361}$$

one can write

$$\vec{F} = [F^\alpha] = \gamma_{v_1}\left[\frac{q(\vec{v}_1 \cdot \vec{E}_e)}{c}, q\vec{E}_e + q\vec{v}_1 \times \vec{B}_m\right]. \tag{2.362}$$

Using the four-velocity

$$\vec{V_1} = [v^b] = \gamma_{v_1}\left[c, \frac{dx^1}{dt}, \frac{dx^2}{dt}, \frac{dx^3}{dt}\right] = \gamma_{v_1}[c, \vec{v_1}], \tag{2.363}$$

we find

$$\vec{V_1} \cdot \vec{F} = v_\alpha f^\alpha = g_{\alpha\beta}v^\beta f^\alpha$$

$$= \gamma_{v_1}[c, -\vec{v_1}] \cdot \gamma_{v_1}\left[\frac{q(\vec{v_1} \cdot \vec{E_e})}{c}, q\vec{E_e} + q\vec{v_1} \times \vec{B_m}\right] \tag{2.364}$$

$$= \gamma_{v_1}^2\left\{q(\vec{v_1} \cdot \vec{E}) - q(\vec{v_1} \cdot \vec{E_e}) - q\vec{v_1} \cdot (\vec{v} \times \vec{B_m})\right\}$$

$$\Rightarrow \vec{V_1} \cdot \vec{F} = 0.$$

This shows that the Lorentz force is a pure force. Consequently, we find

$$\vec{V_1} \cdot \vec{F} = 0$$

$$\Rightarrow \vec{V_1} \cdot \frac{d}{d\tau}(m\vec{V_1}) = m\vec{V_1} \cdot \frac{d\vec{V_1}}{d\tau} + \left(\vec{V_1} \cdot \vec{V_1}\right)\frac{dm}{d\tau}$$

$$= \frac{m}{2}\frac{d}{d\tau}\left(\vec{V_1} \cdot \vec{V_1}\right) + \left(\vec{V_1} \cdot \vec{V_1}\right)\frac{dm}{d\tau} = 0 \tag{2.365}$$

$$\Rightarrow \frac{m}{2}\frac{dc^2}{d\tau} + c^2\frac{dm}{d\tau} = c^2\frac{dm}{d\tau} = 0$$

$$\Rightarrow m = m_0.$$

This shows the mass of the particle does not not change.

Example 2.9. Four-momentum of a photon.

As we saw in the previous example, in the case of a massive object, the proper time τ measured by a clock aboard the object's frame (or by an observer at rest in the IIF) is used as our *affine parameter*. This is because the speed of light, which is independent of the reference frame, gives the equation

$$\left(\frac{ds}{d\tau}\right)^2 = c^2 \Rightarrow s = c\tau + \beta,$$

and in this equation the proper time τ,

$$d\tau = \frac{dt}{\gamma_u} = dt\sqrt{1 - \frac{v_1^2}{c^2}} \Rightarrow \tau = \sqrt{1 - \frac{v_1^2}{c^2}} \int dt, \tag{2.366}$$

can never be zero as the particle travels with a speed $v_1 < c$.

(a) Show that the proper time cannot be an affine parameter for a photon since the worldline for a photon is a null curve.

(b) By considering a photon traveling along the positive x-axis with speed c in the S frame, show that a *privileged family of affine parameter ε*, defined by

$$x^\alpha = u^\alpha \varepsilon, \tag{2.367}$$

where u^α

$$u^\alpha = [1,\ 1,\ 0,\ 0], \tag{2.368}$$

gives a null tangent vector that does not change the geometry of the curve (the photon worldline).

(c) Show that the four-momentum of a photon traveling with speed c at a given direction in the S frame is given by

$$\vec{P} = \left(\frac{E}{c}, \vec{p}\right) = (\hbar k, \hbar \vec{k}),$$

where $\vec{k}$ is the wave vector and k is the wave number, defined by

$$k = \frac{2\pi}{\lambda} \Rightarrow \vec{k} = \left(\frac{2\pi}{\lambda_x}, \frac{2\pi}{\lambda_y}, \frac{2\pi}{\lambda_y}\right), \tag{2.369}$$

and is related to the momentum $\vec{p}$, by the de Broglie wavelength, λ:

$$p = \frac{h}{\lambda} = \frac{h/2\pi}{\lambda/2\pi} = \hbar k \Rightarrow \vec{p} = \hbar \vec{k}. \tag{2.370}$$

Solution:

(a) In the case of a photon or other particles traveling at the speed of light (if it exists), we cannot use the proper time as our affine parameter since for a photon $v_1 = c$:

$$d\tau = \frac{dt}{\gamma_u} = dt\sqrt{1 - \frac{v_1^2}{c^2}} = 0. \tag{2.371}$$

If $\vec{t}$ is the tangent vector defining the worldline of a photon (a curve) in the Minkowski space-time manifold, at any point on the curve an infinitesimal length becomes

$$|\,d\vec{s}\,| = |\,\vec{t}\,|\,d\tau = 0, \tag{2.372}$$

which confirms the worldline of a photon is a *null curve*. For a null curve we cannot use such an affine parameter since it gives a null vector and that does not define the geometry of the curve on the manifold. Therefore, we need to find a different affine parameter.

(b) For a photon traveling along the positive x-axis with speed c in the S frame, for *a privileged family of affine parameter ε* defined by

$$x^\alpha = u^\alpha \varepsilon, \tag{2.373}$$

the coordinates of the photon in the Minkowski space-time are

$$x^0 = ct, \; x^1 = ct, \; x^2 = 0, \; x^3 = 0$$
$$\Rightarrow x^\alpha = [ct, \; ct, \; 0, \; 0] = u^\alpha \varepsilon = [\varepsilon, \; \varepsilon, \; 0, \; 0], \tag{2.374}$$

which means the tangent vector $\vec{u}$ to the curve for the photon in terms of this parameter is given by

$$\vec{u} = \frac{dx^\alpha}{d\varepsilon} \hat{e}_\alpha = u^\alpha \hat{e}_\alpha, \tag{2.375}$$

where

$$u^\alpha = [1, \; 1, \; 0, \; 0], \tag{2.376}$$

which is a null vector, and the curve defined by this tangent vector,

$$\vec{u} \cdot \vec{u} = \left(\frac{ds}{d\varepsilon} \right)^2 = u^\alpha u_\alpha$$

$$\Rightarrow \left(\frac{ds}{d\varepsilon} \right)^2 = u^\alpha g_{\alpha\beta} u^\alpha = [1, \; 1, \; 0, \; 0] \begin{bmatrix} 1 & 0 & 0 & 0 \\ 0 & -1 & 0 & 0 \\ 0 & 0 & -1 & 0 \\ 0 & 0 & 0 & -1 \end{bmatrix} \begin{bmatrix} 1 \\ 1 \\ 0 \\ 0 \end{bmatrix} = 0, \tag{2.377}$$

is still a null curve. This tangent vector, which gives the four-velocity of the photon, is different from that of a massive particle, where we saw

$$\vec{u} \cdot \vec{u} = \left(\frac{ds}{d\tau} \right)^2 = c^2. \tag{2.378}$$

We also note that the equation of motion for a photon can also be expressed in terms of this parameter given by

$$\frac{du^\alpha}{d\varepsilon} e_\alpha = \frac{d}{d\varepsilon}[1, \; 1, \; 0, \; 0] = [0, \; 0, \; 0, \; 0]. \tag{2.379}$$

(c) In the case of a massive particle, to find the four-momentum, we multiplied the four-velocity by the rest mass (proper mass). We follow a similar approach here also. We multiply the four-velocity by some parameter α so that

$$\vec{P} = \alpha \vec{u} \Rightarrow P^\alpha = \alpha V^\alpha. \tag{2.380}$$

Like the massive particle, we represent the energy of the photon by $(\frac{E}{c} = \alpha V^0 = \alpha)$ and the three-momentum by $(\vec{p} = \alpha \vec{v}_1)$ in the S frame. Then the four-momentum of a photon can be expressed as

$$\vec{P} = P^\alpha \hat{e}_\alpha = \left(\frac{E}{c}, \vec{p}\right). \tag{2.381}$$

According to de Broglie, whether massive or non-massive, all particles behave like waves as long as they have momentum:

$$p = \frac{h}{\lambda}. \tag{2.382}$$

The wavelength is often expressed in terms of the magnitude of the wave vector $\vec{k}$, which is defined as

$$k = \frac{2\pi}{\lambda} \Rightarrow \vec{k} = \left(\frac{2\pi}{\lambda_x}, \frac{2\pi}{\lambda_y}, \frac{2\pi}{\lambda_y}\right). \tag{2.383}$$

Then the three-momentum becomes

$$\vec{p} = \frac{h\vec{k}}{2\pi} = \hbar\vec{k}. \tag{2.384}$$

For a photon, the energy is given by

$$E = \frac{hc}{\lambda} = \frac{h}{2\pi}\frac{2\pi}{\lambda}c = \hbar k c, \tag{2.385}$$

and if one express the four-momentum in terms of the four-wave vector,

$$P^\alpha = \frac{hK^\alpha}{2\pi} = \hbar K^\alpha, \tag{2.386}$$

then the four-wave vector $\vec{K}$ can be expressed as

$$\frac{hK^\alpha}{2\pi} = \left(\frac{h}{\lambda}, \frac{h\vec{k}}{2\pi}\right) \Rightarrow \vec{K} = k^\alpha \hat{e}_\alpha = \left(\frac{2\pi}{\lambda}, \vec{k}\right)$$

$$\Rightarrow \hbar K^\alpha = \left(\frac{E}{c}, \vec{p}\right) = \left(\hbar k, \hbar \vec{k}\right) \Rightarrow \vec{K} = \left(k, \vec{k}\right). \tag{2.387}$$

Example 2.10. The Doppler effect.

Let us consider an observer in the S frame defined by the coordinates x^α. This observer receives a photon of wavelength λ at angle θ as measured from the positive x-axis. This photon is emitted by a laser mounted on a spacecraft (S' frame) moving with a velocity v along the positive x-direction as measured by the observer in the S frame (see figure 2.17). We define a set of non-rotating coordinates x'^α on the spacecraft. Applying the four-momentum for a photon in the previous example, find the wavelength λ' of the photon and the angle of emission by the laser θ' on the spacecraft as measured by an observer in the S' frame (the Alien).

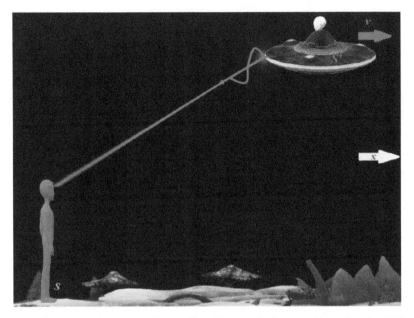

Figure 2.17. A photon emitted by a spaceship traveling with a speed v in the positive x-direction (S' frame). An observer in the S frame measures this photon wavelength to be λ and its direction θ measured from the positive x-axis.

Solution: From example 2.9, we note that the four-wave vector of the photon in the S frame is given by

$$\vec{K} = K^\mu \hat{e}_\mu = \left(\frac{2\pi}{\lambda}, k \cos(\theta), k \sin(\theta), 0 \right) = \frac{2\pi}{\lambda}(1, \cos(\theta), \sin(\theta), 0), \quad (2.388)$$

where we used

$$k = \frac{2\pi}{\lambda}. \qquad (2.389)$$

Note that for the positive x-axis shown in figure 2.17, the z-axis is directed outward normal to the x-axis. In the S' coordinates the four-wave vector is given by

$$\vec{K}' = K'^\nu \hat{e}'_\nu = \left(\frac{2\pi}{\lambda'}, k' \cos(\theta'), k' \sin(\theta'), 0 \right) = \frac{2\pi}{\lambda'}(1, \cos(\theta'), \sin(\theta'), 0). \quad (2.390)$$

We want to find the components K'^ν. These components can easily be determined from the fact that vectors are geometrical properties and would remain the same. That means the four-wave vector must be the same in the S and S' coordinates (i.e., $\vec{K} = \vec{K}'$). Therefore, we can write

$$K'^\alpha = \vec{K} \cdot \hat{e}'^\alpha = K^\beta \hat{e}_\beta \cdot \hat{e}'^\alpha = K^\beta \Lambda'^\alpha_\kappa \hat{e}_\beta \cdot \hat{e}^\kappa = K^\beta \Lambda'^\alpha_\kappa \delta^\kappa_\beta = \Lambda'^\alpha_\beta K^\beta. \qquad (2.391)$$

Using

$$\Lambda^{'\alpha}_{\beta} = \begin{bmatrix} \gamma & -\gamma\beta & 0 & 0 \\ -\gamma\beta & \gamma & 0 & 0 \\ 0 & 0 & 1 & 0 \\ 0 & 0 & 0 & 1 \end{bmatrix}, \tag{2.392}$$

one can write

$$[K'^{\alpha}] = \frac{2\pi}{\lambda}\begin{bmatrix} \gamma & -\gamma\beta & 0 & 0 \\ -\gamma\beta & \gamma & 0 & 0 \\ 0 & 0 & 1 & 0 \\ 0 & 0 & 0 & 1 \end{bmatrix}\begin{bmatrix} 1 \\ \cos(\theta) \\ \sin(\theta) \\ 0 \end{bmatrix}$$

$$= \frac{2\pi}{\lambda}\begin{bmatrix} \gamma - \gamma\beta\cos(\theta) \\ -\gamma\beta + \gamma\cos(\theta) \\ \sin(\theta) \\ 0 \end{bmatrix} \tag{2.393}$$

so that

$$[K'^{\alpha}] = \frac{2\pi}{\lambda'}\begin{bmatrix} 1 \\ \cos(\theta') \\ \sin(\theta') \\ 0 \end{bmatrix} = \frac{2\pi}{\lambda}\begin{bmatrix} \gamma - \gamma\beta\cos(\theta) \\ -\gamma\beta + \gamma\cos(\theta) \\ \sin(\theta) \\ 0 \end{bmatrix}. \tag{2.394}$$

From the first row it follows that

$$\frac{1}{\lambda'} = \frac{\gamma}{\lambda}(1 - \beta\cos(\theta)) \Rightarrow \frac{\lambda}{\lambda'} = \gamma(1 - \beta\cos(\theta)). \tag{2.395}$$

From the second and third rows, we find

$$\frac{2\pi}{\lambda'}\cos(\theta') = \frac{2\pi\gamma}{\lambda}(\cos(\theta) - \beta) \Rightarrow \frac{\lambda}{\lambda'}\cos(\theta') = \gamma(\cos(\theta) - \beta), \tag{2.396}$$

$$\frac{2\pi}{\lambda'}\sin(\theta') = \frac{2\pi}{\lambda}\sin(\theta) \Rightarrow \frac{\lambda}{\lambda'}\sin(\theta') = \sin(\theta). \tag{2.397}$$

Dividing equation (2.397) by equation (2.396), the angle of emission by the laser pointer on the S' frame becomes

$$\frac{\sin(\theta')}{\cos(\theta')} = \frac{\sin(\theta)}{\gamma(\cos(\theta) - \beta)} \Rightarrow \tan(\theta') = \frac{\tan(\theta)}{\gamma[1 - (v/c)\sec(\theta)]}. \tag{2.398}$$

Equation (2.398) is a version of the *relativistic aberration formula*. For $\theta = 0$, equation (2.395) reduces to

$$\frac{\lambda}{\lambda'} = \gamma(1 - \beta) = \frac{1 - \dfrac{v}{c}}{\sqrt{1 - \dfrac{v^2}{c^2}}} = \sqrt{\frac{1 - \dfrac{v}{c}}{1 + \dfrac{v}{c}}}. \tag{2.399}$$

In terms of the frequencies ($\lambda = c/f$), we find

$$\frac{f'}{f} = \sqrt{\frac{1 - \dfrac{v}{c}}{1 + \dfrac{v}{c}}}. \tag{2.400}$$

On the other hand for $\theta = \pi$ equation (2.395) becomes

$$\frac{\lambda}{\lambda'} = \frac{1 + \dfrac{v}{c}}{\sqrt{1 - \dfrac{v^2}{c^2}}} = \sqrt{\frac{1 + \dfrac{v}{c}}{1 - \dfrac{v}{c}}} \Rightarrow \frac{f}{f'} = \sqrt{\frac{1 - \dfrac{v}{c}}{1 + \dfrac{v}{c}}}, \tag{2.401}$$

which is the equation for a *Doppler effect*.

Note: In our derivation of the Doppler effect even though the emission and observation of the photon are two different events that took place at different times, we held the four-momentum to be constant. This is because the equation of motion for a photon in terms of the privileged family of affine parameter ε we used for the worldline is a constant,

$$\frac{d\vec{P}}{d\varepsilon} = \alpha \frac{du^\mu}{d\varepsilon} e_\mu = \alpha \frac{d}{d\varepsilon}[1, 1, 0, 0] = [0, 0, 0, 0]. \tag{2.402}$$

where α is a parameter for the photon like the proper mass for a massive particle, m_0.

Example 2.11. Compton scattering.

The collision of an electron and a photon, known as *Compton scattering*, can be described using relativistic four-momentum. Consider a photon traveling along the positive x-axis with four-momentum $\vec{P}$, which collides with an electron that is at rest at the origin in the S frame that is at rest. Suppose the photon has a frequency ν and the electron has a rest mass m_0 in the S frame. Before the photon collides with the electron, the four-momentum for the photon can be expressed as

$$\vec{P}_{ph} = p_{ph}^\mu \hat{e}_\mu = \left(\frac{h\nu}{c}, \vec{p}_p\right) = \left(\frac{h\nu}{c}, \frac{hk}{2\pi}, 0, 0\right) = \left(\frac{h\nu}{c}, \frac{h\nu}{c}, 0, 0\right)$$
$$\Rightarrow \vec{P}_{ph} = \frac{h\nu}{c}(1, 1, 0, 0), \tag{2.403}$$

where we used

$$k = \frac{2\pi}{\lambda} = \frac{2\pi\nu}{c}. \tag{2.404}$$

For the electron, the three-velocity, $\vec{v}_1 = 0$, the four-momentum before the collision can be expressed as

$$\vec{P}_{el} = P^\mu_{el}\hat{e}_\mu = (m_0 c, \vec{p}_{el}) = m_0 c(1, 0, 0, 0). \tag{2.405}$$

After the collision as shown in figure 2.18, the electron would transfer some of its momentum to the electron. As a result, the photon frequency changes to $\bar{\nu}$ and the electron velocity to $\vec{u}$. Using conservation of the four-momentum shows that the photon frequency after the collision is given by

$$\bar{\nu} = \nu\left[1 + \frac{h\nu}{m_0 c^2}(1 - \cos(\theta))\right]^{-1}. \tag{2.406}$$

Solution: Referring to figure 2.18 for the photon, the four-momentum after the collision becomes

$$\vec{P}_{ph} = p^\mu_{ph}\hat{e}_\mu = \left(\frac{h\bar{\nu}}{c}, \frac{h\bar{\nu}\cos(\theta)}{c}, \frac{h\bar{\nu}\sin(\theta)}{c}, 0\right) = \frac{h\bar{\nu}}{c}(1, \cos(\theta), \sin(\theta), 0), \tag{2.407}$$

and for the electron

$$\vec{P}_{el} = p^\mu_{el}\hat{e}_\mu = \gamma_{v_2}(m_0 c, m_0\vec{v}_2) = \gamma_{v_2}(m_0 c, m_0 v_2\cos(\phi), -m_0 v_2\sin(\phi), 0), \tag{2.408}$$

where

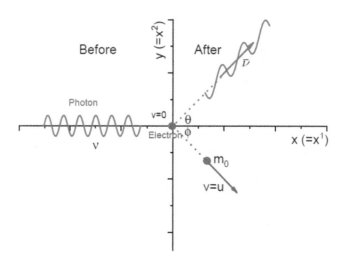

Figure 2.18. Compton scattering.

$$\gamma_{v_2} = \frac{1}{\sqrt{1 - \dfrac{v_2{}^2}{c^2}}}. \tag{2.409}$$

Note that we need to find v_2. The total four-momentum before and after the collision must be the same, which leads to

$$(m_0c, 0, 0, 0) + \frac{h\nu}{c}(1, 1, 0, 0) = \gamma_{v_2}(m_0c, m_0v_2\cos(\phi), -m_0v_2\sin(\phi), 0)$$

$$+ \frac{h\bar{\nu}}{c}(1, \cos(\theta), \sin(\theta), 0)$$

$$\Rightarrow \frac{h\nu}{c}\left(1 + \frac{m_0c^2}{h\nu}, 1, 0, 0\right) \tag{2.410}$$

$$= \frac{h\bar{\nu}}{c}\left(1 + \frac{\gamma_{v_2}m_0c^2}{h\bar{\nu}}, \cos(\theta) + \frac{c}{h\bar{\nu}}\gamma_{v_2}m_0v_2\cos(\phi), \sin(\theta)\right.$$

$$\left. - \frac{c}{h\bar{\nu}}\gamma_{v_2}m_0v_2\sin(\phi), 0\right).$$

From this equation, one finds

$$m_0c + \frac{h\nu}{c} = \gamma_u m_0 c + \frac{h\bar{\nu}}{c} \Rightarrow \gamma_u = 1 + \frac{h}{m_0c^2}(\nu - \bar{\nu}),$$

$$\frac{h\nu}{c} = \frac{h\bar{\nu}}{c}\cos(\theta) + \gamma_u m_0 u \cos(\phi) \Rightarrow \gamma_u m_0 u \cos(\phi) = \frac{h\nu}{c} - \frac{h\bar{\nu}}{c}\cos(\theta) \tag{2.411}$$

$$0 = \sin(\theta) - \frac{c}{h\bar{\nu}}\gamma_u m_0 u \sin(\phi)$$

$$\Rightarrow \frac{c}{h\bar{\nu}}\gamma_u m_0 u \sin(\phi) = \sin(\theta).$$

Combining the last two equations, we have

$$\cos(\phi) = \frac{\nu - \bar{\nu}\cos(\theta)}{\bar{\nu}\sin(\theta)}\sin(\phi) \Rightarrow \sin^2(\phi)\left[1 + \left(\frac{\nu - \bar{\nu}\cos(\theta)}{\bar{\nu}\sin(\theta)}\right)^2\right] = 1$$

$$\Rightarrow \sin^2(\phi) = \frac{1}{1 + \left(\dfrac{\nu - \bar{\nu}\cos(\theta)}{\bar{\nu}\sin(\theta)}\right)^2}. \tag{2.412}$$

Substituting this into equations (2.411) and (2.412), and also using equation (2.409), one can show that

$$\bar{\nu} = \nu\left[1 + \frac{h\nu}{m_0c^2}(1 - \cos(\theta))\right]^{-1}. \tag{2.413}$$

2.16 Parallel transport

In order to understand the idea of parallel transport of a vector on a manifold, let us consider the motion of a particle in space. Suppose the position of the particle depends on time t, then the displacement $\vec{D}$ is parametrized by time t ($\vec{D}(t)$). The velocity of the particle is given by

$$\vec{v} = \frac{d\vec{D}}{dt}, \tag{2.414}$$

and the acceleration

$$\vec{a} = \frac{d\vec{v}}{dt}. \tag{2.415}$$

Suppose you plot the displacement of the particle at different times, then you would get generally the curve $C(t)$ shown in figure 2.19. The particle would have a constant velocity throughout this curve provided its acceleration is zero:

$$\vec{a} = \frac{d\vec{v}}{dt} = 0 \Rightarrow \vec{v} = \text{constant.} \tag{2.416}$$

This means the particle travels along this curve with a constant velocity. The velocity would have the same magnitude and direction. The velocity vector remains parallel at each point on the curve describing the displacement of the particle as a function of time. For the Cartesian coordinates shown in figure 2.19, the vector can be expressed as

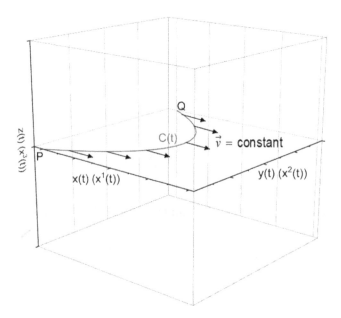

Figure 2.19. Parallel transport of a vector $\vec{v}(t)$ along a curve $C(t)$.

$$\vec{v} = v_1 \hat{x} + v_2 \hat{y}. \tag{2.417}$$

For a parallel transport of this vector along the curve $C(t)$,

$$\frac{d\vec{v}}{dt} = 0 \Rightarrow \frac{dv_1}{dt}\hat{x} + \frac{dv_2}{dt}\hat{y} = 0 \Rightarrow \frac{dv_1}{dt} = \frac{dv_2}{dt} = 0,$$
$$\Rightarrow \frac{dv_\alpha}{dt} = 0 \tag{2.418}$$

as the basis vectors ($\hat{e}^1(t) = \hat{x}$ and $\hat{e}^2(t) = \hat{y}$) are constant throughout the curve and independent of the parameter t. On the other hand, for curved coordinates x^μ, the dual basis vectors $\hat{e}^\mu(x)$ change with the parameter that defines the coordinates. On a manifold defined by the coordinates $x^\mu(u)$ and with dual basis vector $\hat{e}^\mu(x(u))$, a vector

$$\vec{v} = v_\mu(u)\hat{e}^\mu(u) \tag{2.419}$$

can be parallel transported on a curve $C(\mu)$ when its *intrinsic derivative is zero*:

$$\frac{Dv_\mu}{Du} = \frac{dv_\mu}{du} - \Gamma^\beta_{\mu\kappa}v_\beta\frac{dx^\kappa}{du} = 0, \tag{2.420}$$

which we often put in the form

$$\frac{dv_\mu}{du} = \Gamma^\beta_{\mu\kappa}v_\beta\frac{dx^\kappa}{du}. \tag{2.421}$$

We recall that for a local Cartesian coordinates the affine connection $\Gamma^\beta_{\mu\kappa} = 0$ and we find

$$\frac{dv_\mu}{du} = 0, \tag{2.422}$$

which is the requirement for parallel transport of a vector along the curve $C(u)$ like the velocity vector in equation (2.418).

Example 2.12. Show that the four-momentum for a free particle is parallel transported in the Minkowski space-time.

   Solution: For free massive particles, where the four-force is zero, the equation of motion is given by

$$\vec{F} = \frac{d\vec{P}}{d\tau} = \frac{Dp^\kappa}{D\tau}\hat{e}_\kappa = 0, \tag{2.423}$$

where the proper time, τ, is the affine parameter along the particle's worldline. Similarly, for photons, we have

$$\frac{d\vec{P}}{d\epsilon} = 0, \tag{2.424}$$

where ϵ is the affine parameter along the photon worldline. We know that for a parallel transport of a vector on a curve C on a manifold, its intrinsic derivative vanishes. Then for the four-momentum parallel transport, one can write

$$\frac{Dp_\alpha}{Du} = \frac{dp_\alpha}{du} - \Gamma^\beta_{\alpha\kappa} p_\beta \frac{dx^\kappa}{du} = 0 \tag{2.425}$$

In the case of a free particle or a photon in the Minkowski space-time, where the affine connection $\Gamma^\beta_{\alpha\kappa} = 0$ in Cartesian coordinates, the four-momentum becomes

$$\frac{Dp_\alpha}{D\tau} = \frac{dp_\alpha}{d\tau} = 0, \ \frac{Dp_\alpha}{D\epsilon} = \frac{dp_\alpha}{d\epsilon} = 0 \tag{2.426}$$

along the respective worldlines. Thus, in special relativity the worldlines for free particles and photons are parallel transported in Minkowski space-time.

2.17 The geodesic

We recall that a local geodesic at point P is where the affine connection is zero:

$$\Gamma'^\beta_{\mu\kappa}(P) = 0. \tag{2.427}$$

Suppose we have a set of points in Euclidean space defining a geodesic curve, then at all of these points the affine connection must vanish:

$$\partial_\kappa \hat{e}_\beta = \Gamma^f_{\beta\kappa} \hat{e}_f = 0. \tag{2.428}$$

The basis vector with respect to the coordinates x^κ does not change along the geodesic curve. This means for the tangent vector

$$\vec{t} = \frac{dx^\mu}{du} \hat{e}_\mu, \tag{2.429}$$

at least the direction (determined by $\hat{e}_\mu$) remains the same. In Euclidean space this makes the curve become a straight line, where the tangent vectors have the same direction along the line. Thus for Euclidean space the geodesic is a straight line.

For a general curve defined by $x^\mu = x^\mu(u)$ on a manifold, if the curve is geodesic then the tangent vector must have the same direction at all points on the curve. This means the change in the tangent vector with respect to the parameter u then becomes

$$\frac{d\vec{t}}{du} = \lambda(u)\vec{t} = \lambda(u)\frac{dx^\mu}{du}\hat{e}_\mu, \tag{2.430}$$

where $\lambda(u)$ is some function of u. Using our result for the intrinsic derivative of a vector in equation (2.263) for the tangent vector, one can write

$$\frac{d\vec{t}}{du} = \frac{Dt^\mu}{Du}\hat{e}_\mu = \left(\frac{dt^\mu}{du} + \Gamma^\mu_{\kappa\beta} t^\kappa \frac{dx^\beta}{du}\right)\hat{e}_\mu, \tag{2.431}$$

so that substituting this into equation (2.430), we find

$$\frac{dt^\mu}{du} + \Gamma^\mu_{\kappa\beta}t^\kappa\frac{dx^\beta}{du} = \lambda(u)\frac{dx^\mu}{du}. \tag{2.432}$$

Noting that

$$\vec{t} = \frac{dx^\kappa}{du}\hat{e}_\kappa = t^\kappa\hat{e}_\kappa \Rightarrow t^\kappa = \frac{dx^\kappa}{du}, \tag{2.433}$$

equation (2.432) can be written as

$$\frac{d^2x^\mu}{du^2} + \Gamma^\mu_{\kappa\beta}\frac{dx^\kappa}{du}\frac{dx^\beta}{du} = \lambda(u)\frac{dx^\mu}{du}. \tag{2.434}$$

The result in equation (2.434) is valid for both null and non-null geodesics parametrized in terms of some general parameter u. For an *affine parameter u*, where it is related to the distance s on the curve by

$$u = ms + k, \tag{2.435}$$

we have

$$du = mds \tag{2.436}$$

and

$$|\vec{t}| = \left|\frac{ds}{du}\right| = \frac{1}{m} = \text{constant}, \tag{2.437}$$

which is a tangent vector with a constant length that is independent of the parameter u. This means

$$\frac{d\vec{t}}{du} = \frac{d}{du}(|\vec{t}|\hat{t}) = \hat{t}\frac{d}{du}|\vec{t}| = 0$$
$$\Rightarrow \frac{d\vec{t}}{du} = \lambda(u)\vec{t} = 0 \Rightarrow \lambda(u) = 0. \tag{2.438}$$

Therefore, in general for an *affine parameter* the equation for the geodesic in equations (2.432) and (2.434) can be written as

$$\frac{Dt^\mu}{Du} = \frac{dt^\mu}{du} + \Gamma^\mu_{\kappa\beta}t^\kappa\frac{dx^\beta}{du} = 0, \tag{2.439}$$

and

$$\frac{d^2x^\mu}{du^2} + \Gamma^\mu_{\kappa\beta}\frac{dx^\kappa}{du}\frac{dx^\beta}{du} = 0, \tag{2.440}$$

respectively. Equations (2.440) or (2.439) describe a *parallel transport* for the tangent vector, which we discussed in the previous section. Note that the intrinsic derivative for a contravariant component of a vector is

$$\frac{Dv^\mu}{Du} = \frac{dv^\mu}{du} + \Gamma^\mu_{\kappa\beta}v^\kappa\frac{dx^\beta}{du}, \tag{2.441}$$

whereas the covariant component (see problem 7) is

$$\frac{Dv_\mu}{Du} = \frac{dv_\mu}{du} - \Gamma^\beta_{\mu\kappa}v_\beta\frac{dx^\kappa}{du}. \tag{2.442}$$

It can be shown that (see problem 8) if one changes the affine parameter u to u', the coordinates that define the geodesic curve would change from $x^\mu(u)$ to $x^\mu(u')$, and equation (2.440) becomes

$$\frac{d^2x^\mu}{du'^2} + \Gamma^\mu_{\kappa\beta}\frac{dx^\kappa}{du'}\frac{dx^\beta}{du'} = \left(\frac{\dfrac{d^2u}{du'^2}}{\dfrac{du}{du'}}\right)\frac{dx^\mu}{du'}. \tag{2.443}$$

We will see how this relation can be derived from the Euler–Lagrange equation.

2.18 The Euler–Lagrange equation

In a Euclidean space a surface is defined by the function $F(x, y, z)$, where it depends on the Cartesian coordinates x, y, and z (see figure 2.20(a)). Instead of a Euclidean space, let us consider a surface defined by the function $F\left(x, y(x), y'(x) = \frac{dy}{dx}\right)$. This surface could, for example, be a surface in phase space (in classical mechanics) if we replace

$$x \to t, \; y(x) \to y(t), \; y'(x) \to y'(t) = \frac{dy}{dt} = v_y(t) = \frac{p_y}{m},$$

describing the dynamics of a particle mass, m, moving along the y-direction in terms of the parameters $\left(\text{time} = t, \text{ position} = y(t), \text{ velocity} = v_y(t)\hat{y} = \frac{p_y\hat{y}}{m}\right)$, where $p_y\hat{y}$ is

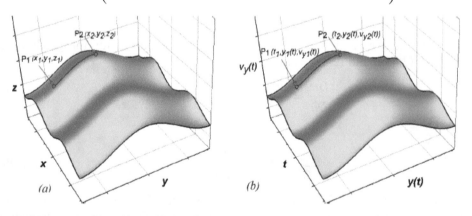

Figure 2.20. (a) A surface defined by the function $F(x, y, z)$ in Euclidian space and (b) a surface defined by the function $F\left(t, y(t), v_y(t) = \frac{p_y(t)}{m}\right)$ in a phase space.

the momentum (see figure 2.20(b)). In classical mechanics, the dynamics of a particle is determined by an equation derived from Newton's second law. This equation can be derived from a more general equation known as *the Euler–Lagrange equation*:

$$\frac{\partial}{\partial x}\left(\frac{\partial F}{\partial y'}\right) - \frac{\partial F}{\partial y} = 0, \tag{2.444}$$

where $F = F(x, y(x), y'(x))$ is the function that defines the surface constructed by the set of points with coordinates $(x, y(x), y'(x))$. The Euler–Lagrange equation is derived by *applying the calculus of variations*. In general, in the problem that we want to solve applying the calculus of variations, we know the coordinates of two different points $(x_1, y(x_1), y'(x_1))$ and $(x_2, y(x_2), y'(x_2))$ *on the surface defined by* $F = F(x, y(x), y'(x))$. From the infinitely many trajectories that can connect these two points, there is only one trajectory on this surface that is the shortest (the Geodesic). Finding such Geodesic is the most general problem that can be solved by applying the calculus of variations.

The surface is defined by the function $F(x, y(x), y'(x))$. The distance between these two points is determined by evaluating the integral:

$$I = \int_{x_1}^{x_2} F(x, y(x), y'(x))\, dx. \tag{2.445}$$

To determine the equation that the function F is governed by, so that we can find the shortest length joining the two points (see figure 2.21), let the function for any path connecting the two points be $Y(x)$. From these infinite number of functions, only one function gives the minimum distance between the two points. If this function is $y(x)$, then we may write $Y(x)$ in terms of $y(x)$ as

$$Y(x, \epsilon) = y(x) + \epsilon\eta(x), \tag{2.446}$$

where $\eta(x)$ is an arbitrary function that must satisfy the condition

$$\eta(x_1) = \eta(x_2) = 0, \tag{2.447}$$

so that at the two points $(P_1 = (x_1, y_1)$ and $P_2 = (x_2, y_2))$, we find

$$Y(x_1, \epsilon) = y(x_1) = y_1, \; Y(x_2, \epsilon) = y(x_2) = y_2. \tag{2.448}$$

Using equation (2.446), we can also establish the relations

$$Y(x, \epsilon)\,|_{\epsilon=0} = y(x), \tag{2.449}$$

$$\frac{dY(x, \epsilon)}{d\epsilon} = \eta(x) \Rightarrow \frac{dY(\epsilon)}{d\epsilon}\bigg|_{\epsilon=0} = \eta(x), \tag{2.450}$$

$$\frac{dY}{dx} = \frac{dy}{dx} + \epsilon\frac{d\eta}{dx} \text{ or } Y'(x, \epsilon) = y'(x) + \epsilon\eta'(x) \Rightarrow Y'(x, \epsilon)\,|_{\epsilon=0} = y'(x), \tag{2.451}$$

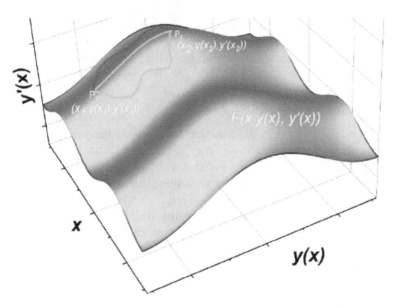

Figure 2.21. A surface defined by the function $F(x, y(x), y'(x) = \frac{dy}{dx})$.

$$\frac{dY'(\epsilon)}{d\epsilon} = \frac{d}{d\epsilon}[y'(x) + \epsilon\eta'(x)] = \eta'(x) \Rightarrow \frac{dY'(\epsilon)}{d\epsilon}\bigg|_{\epsilon=0} = \eta'(x). \qquad (2.452)$$

For the geodesic the integral

$$I(\epsilon) = \int_{x_1}^{x_2} F(x, \, Y(x, \, \epsilon), \, Y'(x, \, \epsilon)) \, dx, \qquad (2.453)$$

must be stationary, which means

$$\frac{dI(\epsilon)}{d\epsilon}\bigg|_{\epsilon=0} = \int_{x_1}^{x_2} \frac{d}{d\epsilon}[F(x, \, Y(x, \, \epsilon), \, Y'(x, \, \epsilon))]\bigg|_{\epsilon=0} dx = 0. \qquad (2.454)$$

Noting that

$$\frac{d}{d\epsilon}[F(x, \, Y(x, \, \epsilon), \, Y'(x, \, \epsilon))]\bigg|_{\epsilon=0} = \frac{\partial F}{\partial Y}\frac{dY(\epsilon)}{d\epsilon} + \frac{\partial F}{\partial Y'}\frac{dY'(\epsilon)}{d\epsilon}\bigg|_{\epsilon=0}$$

$$= \frac{\partial F}{\partial Y}\bigg|_{\epsilon=0}\frac{dY(\epsilon)}{d\epsilon}\bigg|_{\epsilon=0} + \frac{\partial F}{\partial Y'}\bigg|_{\epsilon=0}\frac{dY'(\epsilon)}{d\epsilon}\bigg|_{\epsilon=0}, \qquad (2.455)$$

and using equations (2.449)–(2.452), we find

$$\frac{d}{d\epsilon}[F(x, \, Y(x, \, \epsilon), \, Y'(x, \, \epsilon))]\bigg|_{\epsilon=0} = \left[\frac{\partial}{\partial y}F(x, \, y(x), \, y'(x))\right]\eta(x)$$

$$+ \left[\frac{\partial}{\partial y'}F(x, \, y(x), \, y'(x))\right]\eta'(x). \qquad (2.456)$$

Then in view of equation (2.456), the integral for the geodesic line in equation (2.454) becomes

$$\left.\frac{dI(\epsilon)}{d\epsilon}\right|_{\epsilon=0} = \int_{x_1}^{x_2} \left[\frac{\partial}{\partial y}F(x, y(x), y'(x))\right]\eta(x)dx$$

$$+ \int_{x_1}^{x_2} \left[\frac{\partial}{\partial y'}F(x, y(x), y'(x))\right]\eta'(x)dx. \tag{2.457}$$

Using integration by parts the second integral can be rewritten as

$$\int_{x_1}^{x_2} \left[\frac{\partial}{\partial y'}F(x, y(x), y'(x))\right]\eta'(x)dx = \eta(x)\frac{\partial}{\partial x}\left[\frac{\partial F}{\partial y'}\right]\Bigg|_{x_1}^{x_2}$$

$$- \int_{x_1}^{x_2} \eta(x)\frac{\partial}{\partial x}\left[\frac{\partial F}{\partial y'}\right]\eta(x)dx, \tag{2.458}$$

so that using equation (2.447), we find

$$\int_{x_1}^{x_2} \left[\frac{\partial}{\partial y'}F(x, y(x), y'(x))\right]\eta'(x)dx = -\int_{x_1}^{x_2} \eta(x)\frac{\partial}{\partial x}\left[\frac{\partial F}{\partial y'}\right]\eta(x)dx. \tag{2.459}$$

In view of the result in equation (2.459), equation (2.457) can be put in the form

$$\left.\frac{dI(\epsilon)}{d\epsilon}\right|_{\epsilon=0} = \int_{x_1}^{x_2} \left[\frac{\partial}{\partial y}F(x, y(x), y'(x))\right]\eta(x)dx +$$

$$- \int_{x_1}^{x_2} \eta(x)\frac{\partial}{\partial x}\left(\frac{\partial F}{\partial y'}\right)\eta(x)dx = 0. \tag{2.460}$$

$$\Rightarrow \left.\frac{dI(\epsilon)}{d\epsilon}\right|_{\epsilon=0} = \int_{x_1}^{x_2} \left[\frac{\partial}{\partial x}\left(\frac{\partial F}{\partial y'}\right) - \frac{\partial F}{\partial y}\right]\eta(x)dx = 0.$$

Then it follows that

$$\frac{\partial}{\partial x}\left(\frac{\partial F}{\partial y'}\right) - \frac{\partial F}{\partial y} = 0, \tag{2.461}$$

where $F = F(x, y(x), y'(x))$. Equation (2.461) is known as the Euler–Lagrange equation, which makes the integral in equation (2.445) stationary.

We next change the notation and re-derive the Euler–Lagrange equation and, in the next section, we show the Euler–Lagrange equation is nothing but *the geodesic in equation (2.443)*. To this end, we represent $(x, y(x), y'(x))$ by $\left(u', x^\mu(u'), \frac{dx^\mu(u')}{du'}\right)$ so that the surface is defined by the function $F\left(u', x^\mu(u'), \frac{dx^\mu(u')}{du'}\right)$. The distance between these two points is determined by evaluating the integral:

$$I = \int_{x_1^\mu}^{x_2^\mu} F\left(u', x^\mu(u'), \frac{dx^\mu(u')}{du'}\right) du', \tag{2.462}$$

where x_1^μ and x_2^μ are the coordinates of the two points on the surface. To determine the equation that the function F is governed by so that we find the shortest length joining the two points, let the function for any path connecting the two points be $X^\mu(u')$. From these infinite number of functions there is only one function that gives the minimum distance between the two points. If this particular function is $x^\mu(u')$, then we may write

$$X^\mu(u', \epsilon) = x^\mu(u') + \epsilon\eta(x^\mu), \tag{2.463}$$

where $\eta(x^\mu)$ is an arbitrary function that must satisfy the condition

$$\eta(x_1^\mu) = \eta(x_2^\mu) = 0, \tag{2.464}$$

since all the arbitrary functions for the different paths connecting the two points on the surface, including the particular function $x^\mu(u')$, have the same value. From equation (2.463), we also have

$$\frac{dX^\mu(u', \epsilon)}{d\epsilon} = \eta(x^\mu) \Rightarrow \frac{dX^\mu(u', \epsilon)}{d\epsilon}\bigg|_{\epsilon=0} = \eta(x^\mu), \tag{2.465}$$

and

$$\frac{dX^\mu(u', \epsilon)}{du'} = \frac{dx^\mu(u')}{du'} + \epsilon\frac{d\eta(x^\mu)}{du'}, \Rightarrow \frac{dX^\mu(u', \epsilon)}{du'}\bigg|_{\epsilon=0} = \frac{dx^\mu(u')}{du'},$$
$$\frac{d}{d\epsilon}\left[\frac{dX^\mu(u', \epsilon)}{du'}\right] = \frac{d}{d\epsilon}\left[\frac{dx^\mu(u')}{du'} + \epsilon\frac{d\eta(x^\mu)}{du'}\right] = \frac{d\eta(x^\mu)}{du'} \tag{2.466}$$
$$\Rightarrow \frac{d}{d\epsilon}\left[\frac{dX^\mu(u', \epsilon)}{du'}\right]\bigg|_{\epsilon=0} = \frac{d\eta(x^\mu)}{du'}.$$

For the geodesic (the shortest path defined by $x^\mu(u')$), the integral

$$I(\epsilon) = \int_{x_1^\mu}^{x_2^\mu} F(u', X^\mu(u', \epsilon), \frac{dX^\mu(u', \epsilon)}{du'}) du' \tag{2.467}$$

must be stationary, and

$$\frac{dI(\epsilon)}{d\epsilon}\bigg|_{\epsilon=0} = \int_{x_1^\mu}^{x_2^\mu} \frac{d}{d\epsilon}[F(u', X^\mu(u', \epsilon), \dot{X}^\mu(u', \epsilon))]\bigg|_{\epsilon=0} du' = 0, \tag{2.468}$$

where we introduced the notation

$$\frac{dX^\mu(u', \epsilon)}{du'} = \dot{X}^\mu(u', \epsilon). \tag{2.469}$$

Upon carrying out the differentiating in equation (2.468), we find

$$\int_{x_1^\mu}^{x_2^\mu} \left[\frac{\partial F(u', X^\mu(u', \epsilon), \dot{X}^\mu(u', \epsilon))}{\partial X^\mu} \bigg|_{\epsilon=0} \frac{dX^\mu(u', \epsilon)}{d\epsilon} \bigg|_{\epsilon=0} \right.$$
$$\left. + \frac{\partial F(u', X^\mu(u', \epsilon), \dot{X}^\mu(u', \epsilon))}{\partial \dot{X}^\mu} \bigg|_{\epsilon=0} \frac{d\dot{X}^\mu}{d\epsilon} \bigg|_{\epsilon=0} \right] du' = 0,$$

(2.470)

so that, using the results in equations (2.465) and (2.466), and noting that

$$\frac{\partial F(u', X^\mu(u', \epsilon), \dot{X}^\mu(u', \epsilon))}{\partial X^\mu} \bigg|_{\epsilon=0} = \frac{\partial F(u', x^\mu(u'), \dot{x}^\mu(u'))}{\partial x^\mu},$$
$$\frac{\partial F(u', X^\mu(u', \epsilon), \dot{X}^\mu(u', \epsilon))}{\partial \dot{X}^\mu} \bigg|_{\epsilon=0} = \frac{\partial F(u', x^\mu(u'), \dot{x}^\mu(u'))}{\partial \dot{x}^\mu},$$

(2.471)

we find

$$\int_{x_1^\mu}^{x_2^\mu} \left[\frac{\partial}{\partial x^\mu} F(u', x^\mu(u'), \dot{x}^\mu(u')) \right] \eta(x^\mu) du'$$
$$+ \int_{x_1^\mu}^{x_2^\mu} \left[\frac{\partial}{\partial \dot{x}^\mu} F(u', x^\mu(u'), \dot{x}^\mu(u')) \right] \frac{d\eta(x^\mu)}{du'} du' = 0.$$

(2.472)

Using integration by parts, the second integral can be rewritten as

$$\int_{x_1^\mu}^{x_2^\mu} \left[\frac{\partial}{\partial \dot{x}^\mu} F(u', x^\mu(u'), \dot{x}^\mu(u')) \right] \frac{d\eta(x^\mu)}{du'} du'$$
$$= \eta(x_2^\mu) \left\{ \frac{d}{du'} \left[\frac{\partial}{\partial \dot{x}^\mu} F(u', x^\mu(u'), \dot{x}^\mu(u')) \right] \right\}^{x_2^\mu}$$
$$- \eta(x_1^\mu) \left\{ \frac{d}{du'} \left[\frac{\partial}{\partial \dot{x}^\mu} F(u', x^\mu(u'), \dot{x}^\mu(u')) \right] \right\}_{x_1^\mu}$$
$$- \int_{x_1^\mu}^{x_2^\mu} \eta(x^\mu) \frac{d}{du'} \left[\frac{\partial}{\partial \dot{x}^\mu} F(u', x^\mu(u'), \dot{x}^\mu(u')) \right] du',$$
$$\Rightarrow \int_{x_1^\mu}^{x_2^\mu} \left[\frac{\partial}{\partial \dot{x}^\mu} F(u', x^\mu(u'), \dot{x}^\mu(u')) \right] \frac{d\eta(x^\mu)}{du'} du'$$
$$= - \int_{x_1^\mu}^{x_2^\mu} \eta(x^\mu) \frac{d}{du'} \left[\frac{\partial}{\partial \dot{x}^\mu} F(u', x^\mu(u'), \dot{x}^\mu(u')) \right] du',$$

(2.473)

where we used the relation in equation (2.464). Now, substituting equation (2.473) into equation (2.472), we find

$$= \int_{x_1^\mu}^{x_2^\mu} \left[\frac{d}{du'} \left(\frac{\partial F}{\partial \dot{x}^\mu} \right) - \frac{\partial F}{\partial x^\mu} \right] \eta(x^\mu) du' = 0.$$

(2.474)

From this equation it follows that

$$\frac{d}{du'}\left(\frac{\partial F}{\partial \dot{x}^\mu}\right) - \frac{\partial F}{\partial x^\mu} = 0, \tag{2.475}$$

which is the Euler–Lagrange equation.

2.19 Stationary property of the non-null geodesic

Consider the curve C in our manifold defined by the coordinates $x^\mu(u')$. Suppose we have two points, 1 and 2, on this curve with coordinates x_1^κ and x_2^κ, and we are interested in the length along this curve joining these two points. This length can be determined from

$$L = \int_1^2 ds = \int_1^2 \sqrt{\left|\, g_{\mu\beta}dx^\mu dx^\beta \,\right|} = \int_1^2 \sqrt{\left|\, g_{\mu\beta}\frac{dx^\mu}{du'}\frac{dx^\beta}{du'} \,\right|}\, du', \tag{2.476}$$

or using the notation

$$\frac{dx^\mu}{du'} = \dot{x}^\mu, \tag{2.477}$$

this length can be expressed as

$$L = \int_1^2 \sqrt{\left|\, g_{\mu\beta}\dot{x}^\mu \dot{x}^\beta \,\right|}\, du'. = \int_1^2 F(u', x^\kappa(u'), \dot{x}^\kappa(u'))du', \tag{2.478}$$

where

$$F(u', x^\kappa(u'), \dot{x}^\kappa(u')) = \sqrt{\left|\, g_{\mu\beta}\dot{x}^\mu \dot{x}^\beta \,\right|} = \dot{s} = \frac{ds}{du'}. \tag{2.479}$$

In the previous section, using the calculus of variation for the integral in equation (2.478), if the length is the geodesic (the shortest path) the integral must be stationary and the function must satisfy the *Euler–Lagrange equation*:

$$\frac{d}{du'}\left(\frac{\partial F}{\partial \dot{x}^\kappa}\right) - \frac{\partial F}{\partial x^\kappa} = 0. \tag{2.480}$$

Next, we shall see how the Euler–Lagrange equation is actually the geodesic equation in equation (2.443). To this end, using equation (2.479), we can write

$$\frac{\partial F}{\partial x^\kappa} = \frac{\partial}{\partial x^\kappa}\left[\sqrt{\left|\, g_{\mu\beta}\dot{x}^\mu \dot{x}^\beta \,\right|}\right] = \frac{\dot{x}^\mu \dot{x}^\beta \partial_\kappa g_{\mu\beta}}{2\sqrt{\left|\, g_{\mu\beta}\dot{x}^\mu \dot{x}^\beta \,\right|}} = \frac{\dot{x}^\mu \dot{x}^\beta \partial_\kappa g_{\mu\beta}}{2\dot{s}}, \tag{2.481}$$

and

$$\frac{\partial F}{\partial \dot{x}^\kappa} = \frac{\partial}{\partial \dot{x}^\kappa}\left[\sqrt{\left|\,g_{\mu\beta}\dot{x}^\mu\dot{x}^\beta\,\right|}\right] = \frac{g_{\mu\beta}}{2\sqrt{\left|\,g_{\mu\beta}\dot{x}^\mu\dot{x}^\beta\,\right|}}\left\{\dot{x}^\beta\frac{\partial \dot{x}^\mu}{\partial \dot{x}^\kappa} + \dot{x}^\mu\frac{\partial \dot{x}^\beta}{\partial \dot{x}^\kappa}\right\}$$

$$= \frac{g_{\mu\beta}\dot{x}^\beta}{2\sqrt{\left|\,g_{\mu\beta}\dot{x}^\mu\dot{x}^\beta\,\right|}}\delta_\kappa^\mu + \frac{g_{\mu\beta}\dot{x}^\mu}{2\sqrt{\left|\,g_{\mu\beta}\dot{x}^\mu\dot{x}^\beta\,\right|}}\delta_\kappa^\beta = \frac{g_{\kappa\beta}\dot{x}^\beta}{2\sqrt{\left|\,g_{\kappa\beta}\dot{x}^\kappa\dot{x}^\beta\,\right|}} + \frac{g_{\mu\kappa}\dot{x}^\mu}{2\sqrt{\left|\,g_{\mu\kappa}\dot{x}^\mu\dot{x}^\kappa\,\right|}} \tag{2.482}$$

$$\Rightarrow \frac{\partial F}{\partial \dot{x}^\kappa} = \frac{g_{\kappa\beta}\dot{x}^\beta + g_{\mu\kappa}\dot{x}^\mu}{2\dot{s}},$$

where we replaced

$$\dot{s} = \sqrt{\left|\,g_{\kappa\beta}\dot{x}^\kappa\dot{x}^\beta\,\right|} = \sqrt{\left|\,g_{\mu\kappa}\dot{x}^\mu\dot{x}^\kappa\,\right|}. \tag{2.483}$$

Replacing β by μ in the first term in equation (2.482), and recalling that $g_{\mu\kappa} = g_{\kappa\mu}$, one can write

$$\frac{\partial F}{\partial \dot{x}^\kappa} = \frac{g_{\mu\kappa}\dot{x}^\mu}{\dot{s}}. \tag{2.484}$$

Substituting equations (2.481) and (2.484) into equation (2.480), the Euler–Lagrange equation becomes

$$\frac{d}{du'}\left(\frac{g_{\mu\kappa}\dot{x}^\mu}{\dot{s}}\right) - \frac{\dot{x}^\mu\dot{x}^\beta\partial_\kappa g_{\mu\beta}}{2\dot{s}} = 0. \tag{2.485}$$

For the first term we may write

$$\frac{d}{du'}\left(\frac{g_{\mu\kappa}\dot{x}^\mu}{\dot{s}}\right) = \frac{\dot{x}^\mu}{\dot{s}}\frac{dg_{\mu\kappa}}{du'} + \frac{g_{\mu\kappa}}{\dot{s}}\frac{d\dot{x}^\mu}{du'} + g_{\mu\kappa}\dot{x}^\mu\frac{d}{du'}\left(\frac{1}{\dot{s}}\right)$$

$$= \frac{\dot{x}^\mu}{\dot{s}}\frac{dg_{\mu\kappa}}{du'} + \frac{g_{\mu\kappa}}{\dot{s}}\frac{d\dot{x}^\mu}{du'} - g_{\mu\kappa}\dot{x}^\mu\frac{\ddot{s}}{\dot{s}^2}, \tag{2.486}$$

and noting that

$$\frac{dg_{\mu\kappa}}{du'} = \frac{\partial g_{\mu\kappa}}{\partial x^\beta}\frac{dx^\beta}{du'} = \left(\partial_\beta g_{\mu\kappa}\right)\dot{x}^\beta, \tag{2.487}$$

we find

$$\frac{d}{du'}\left(\frac{g_{\mu\kappa}\dot{x}^\mu}{\dot{s}}\right) = \frac{1}{\dot{s}}\left[\left(\partial_\beta g_{\mu\kappa}\right)\dot{x}^\mu\dot{x}^\beta + g_{\mu\kappa}\ddot{x}^\mu - g_{\mu\kappa}\frac{\ddot{s}}{\dot{s}}\dot{x}^\mu\right]. \tag{2.488}$$

Introducing equation (2.488) into equation (2.485), we find

$$\frac{1}{\dot{s}}\left[\left(\partial_\beta g_{\mu\kappa}\right)\dot{x}^\mu\dot{x}^\beta + g_{\mu\kappa}\ddot{x}^\mu - g_{\mu\kappa}\frac{\ddot{s}}{\dot{s}}\dot{x}^\mu - \frac{\dot{x}^\mu\dot{x}^\beta\partial_\kappa g_{\mu\beta}}{2}\right] = 0.$$

$$\Rightarrow g_{\mu\kappa}\ddot{x}^\mu + \left(\partial_\beta g_{\mu\kappa}\right)\dot{x}^\mu\dot{x}^\beta - \frac{\dot{x}^\mu\dot{x}^\beta\partial_\kappa g_{\mu\beta}}{2} = \frac{\ddot{s}}{\dot{s}}g_{\mu\kappa}\dot{x}^\mu. \tag{2.489}$$

Noting that

$$\left(\partial_\beta g_{\mu\kappa}\right)\dot{x}^\mu\dot{x}^\beta = \left(\partial_\mu g_{\beta\kappa}\right)\dot{x}^\beta\dot{x}^\mu, \tag{2.490}$$

one can write

$$\left(\partial_\beta g_{\mu\kappa}\right)\dot{x}^\mu\dot{x}^\beta = \frac{1}{2}\left[\left(\partial_\beta g_{\mu\kappa}\right)\dot{x}^\mu\dot{x}^\beta + \left(\partial_\mu g_{\beta\kappa}\right)\dot{x}^\beta\dot{x}^\mu\right]. \tag{2.491}$$

Upon substituting equation (2.491) into equation (2.489), we find

$$g_{\mu\kappa}\ddot{x}^\mu + \frac{1}{2}\left[\dot{x}^\mu\dot{x}^\beta\partial_\beta g_{\mu\kappa} + \dot{x}^\beta\dot{x}^\mu\partial_\mu g_{\beta\kappa} - \dot{x}^\mu\dot{x}^\beta\partial_\kappa g_{\mu\beta}\right] = \frac{\ddot{s}}{\dot{s}}g_{\mu\kappa}\dot{x}^\mu$$

$$\Rightarrow g_{\mu\kappa}\ddot{x}^\mu + \frac{1}{2}\dot{x}^\mu\dot{x}^\beta\left[\partial_\beta g_{\mu\kappa} + \partial_\mu g_{\beta\kappa} - \partial_\kappa g_{\mu\beta}\right] = \frac{\ddot{s}}{\dot{s}}g_{\mu\kappa}\dot{x}^\mu \tag{2.492}$$

and, multiplying by $g^{\nu\kappa}$,

$$g^{\nu\kappa}g_{\mu\kappa}\ddot{x}^\mu + \frac{1}{2}g^{\nu\kappa}\left[\partial_\beta g_{\mu\kappa} + \partial_\mu g_{\beta\kappa} - \partial_\kappa g_{\mu\beta}\right]\dot{x}^\mu\dot{x}^\beta = \frac{\ddot{s}}{\dot{s}}g^{\nu\kappa}g_{\mu\kappa}\dot{x}^\mu. \tag{2.493}$$

Now, applying the relation

$$g^{\nu\kappa}g_{\mu\kappa} = \delta_\mu^\nu, \tag{2.494}$$

we find

$$\delta_\mu^\nu\ddot{x}^\mu + \frac{1}{2}g^{\nu\kappa}\left[\partial_\beta g_{\mu\kappa} + \partial_\mu g_{\beta\kappa} - \partial_\kappa g_{\mu\beta}\right]\dot{x}^\mu\dot{x}^\beta = \frac{\ddot{s}}{\dot{s}}\delta_\mu^\nu\dot{x}^\mu. \tag{2.495}$$

which simplifies into

$$\ddot{x}^\nu + \frac{1}{2}g^{\nu\kappa}\left[\partial_\beta g_{\mu\kappa} + \partial_\mu g_{\beta\kappa} - \partial_\kappa g_{\mu\beta}\right]\dot{x}^\mu\dot{x}^\beta = \frac{\ddot{s}}{\dot{s}}\dot{x}^\nu. \tag{2.496}$$

Using the expression for the affine connection in terms of the metric tensor

$$g^{\nu\kappa}\left[\partial_\beta g_{\mu\kappa} + \partial_\mu g_{\beta\kappa} - \partial_\kappa g_{\mu\beta}\right] = \Gamma_{\mu\beta}^\nu, \tag{2.497}$$

equation (2.496) can be put in the form

$$\ddot{x}^\nu + \Gamma_{\mu\beta}^\nu\dot{x}^\mu\dot{x}^\beta = \frac{\ddot{s}}{\dot{s}}\dot{x}^\nu$$

$$\Rightarrow \frac{d^2x^\nu}{du'^2} + \Gamma_{\mu\beta}^\nu\frac{dx^\mu}{du'}\frac{dx^\beta}{du'} = \left(\frac{\dfrac{d^2u}{du'^2}}{\dfrac{du}{du'}}\right)\frac{dx^\nu}{du'}. \tag{2.498}$$

Comparing the result in equation (2.498) with equation (2.443), we can see that the Euler–Lagrange is nothing but *the geodesic* in equation (2.443).

2.20 Homework assignments

Problem 1. Consider the three-dimensional sphere of radius a in a four-dimensional Euclidean space. A point P on this three-dimensional sphere is described by the vector

$$\vec{s} = r \sin(\theta) \cos(\varphi)\hat{x} + r \sin(\theta) \sin(\varphi)\hat{y} + r \cos(\theta)\hat{z} + \sqrt{a^2 - r^2}\,\hat{w}.$$

(a) Find the basis vectors $\hat{e}_r$, $\hat{e}_\theta$, and $\hat{e}_\varphi$ in the tangent space at point P.
(b) Re-derive the metric elements for a three-dimensional sphere from the basis vectors.

Problem 2. Show that for the coordinate transformation x^μ to x'^μ
(a) The dual basis vector

$$\hat{e}'^\mu = \frac{\partial x'^\mu}{\partial x^\kappa}\hat{e}^\kappa.$$

(b) The covariant components of a vector

$$v'_\beta = \frac{\partial x^\mu}{\partial x'^\beta}v_\mu.$$

(c) In example 1.1 we have shown that the basis vectors in the x^μ and x'^μ coordinates for a two-dimensional sphere are given by

$$\hat{e}_1(x^1, x^2) = a \cos(x^1)[\cos(x^2)\hat{x} + \sin(x^2)\hat{y}] - a \sin(x^1)\hat{z},$$
$$\hat{e}_2(x^1, x^2) = a \sin(x^1)[\sin(x^2)\hat{x} - \cos(x^2)\hat{y}]$$

and

$$\hat{e}'_1 = \hat{x} \mp \frac{x'^1}{\sqrt{a^2 - (x'^1)^2 - (x'^2)^2}}\hat{z},$$

$$\hat{e}'_2 = \hat{y} \mp \frac{x'^2}{\sqrt{a^2 - (x'^1)^2 - (x'^2)^2}}\hat{z}.$$

Using the relation in part (a), show that

$$\hat{e}_2(x^1, x^2) = a \sin(x^1)[\sin(x^2)\hat{x} - \cos(x^2)\hat{y}],$$

where the coordinate transformation is defined by

$$x'^1 = a \sin(x^1) \cos(x^2), \quad x'^2 = a \sin(x^1) \sin(x^2),$$

and for a two-dimensional sphere

$$x'^3 = \pm\sqrt{a^2 - (x'^1)^2 - (x'^2)^2} = a\cos(x^1).$$

Problem 3.

(a) Find all the elements for the affine connection, $\Gamma^{\beta}_{\mu\kappa}$, for a point on a two-dimensional sphere embedded in a three-dimensional Euclidean space. Note that in the expressions

$$\frac{\partial\hat{e}_\theta}{\partial\theta} = \Gamma^{\theta}_{\theta\theta}\hat{e}_\theta + \Gamma^{\varphi}_{\theta\theta}\hat{e}_\varphi, \quad \frac{\partial\hat{e}_\theta}{\partial\varphi} = \Gamma^{\theta}_{\theta\varphi}\hat{e}_\theta + \Gamma^{\varphi}_{\theta\varphi}\hat{e}_\varphi,$$

$$\frac{\partial\hat{e}_\varphi}{\partial\theta} = \Gamma^{\theta}_{\varphi\theta}\hat{e}_\theta + \Gamma^{\varphi}_{\varphi\theta}\hat{e}_\varphi, \quad \frac{\partial\hat{e}_\varphi}{\partial\varphi} = \Gamma^{\theta}_{\varphi\varphi}\hat{e}_\theta + \Gamma^{\varphi}_{\varphi\varphi}\hat{e}_\varphi,$$

you are going to determine

$$\Gamma^{\theta}_{\theta\theta}, \Gamma^{\varphi}_{\theta\theta}, \Gamma^{\theta}_{\theta\varphi}, \Gamma^{\varphi}_{\theta\varphi}, \Gamma^{\theta}_{\varphi\theta}, \Gamma^{\varphi}_{\varphi\theta}, \Gamma^{\theta}_{\varphi\varphi}, \text{ and } \Gamma^{\varphi}_{\varphi\varphi}.$$

Also in this case the origin is at the center of the sphere.

(b) Find all the elements for the affine connection, $\Gamma^{\beta}_{\mu\kappa}$, for a point on a three-dimensional sphere embedded in a four-dimensional Euclidean space.

Problem 4. Show that the affine connection under coordinate transformation

$$\Gamma'^{\beta}_{\alpha\mu} = \frac{\partial x'^\beta}{\partial x^\nu}\frac{\partial x^\sigma}{\partial x'^\alpha}\frac{\partial x^\kappa}{\partial x'^\mu}\Gamma^{\nu}_{\sigma\kappa} + \frac{\partial x'^\beta}{\partial x^\nu}\frac{\partial^2 x^\nu}{\partial x'^\mu \partial x'^\alpha},$$

can be written as

$$\Gamma'^{\beta}_{\alpha\mu} = \frac{\partial x'^\beta}{\partial x^\nu}\frac{\partial x^\sigma}{\partial x'^\alpha}\frac{\partial x^\kappa}{\partial x'^\mu}\Gamma^{\nu}_{\sigma\kappa} - \frac{\partial x^\nu}{\partial x'^\alpha}\frac{\partial x^\sigma}{\partial x'^\mu}\frac{\partial^2 x'^\beta}{\partial x^\nu \partial x^\sigma}.$$

Problem 5. Using the coordinate basis vector derived in example 2.1 derive the elements for the metric tensor in the Cartesian coordinates $(x, y) \rightarrow (x^1, x^2)$. You should get

$$g_{11} = \frac{a^2 - (x'^2)^2}{a^2 - (x'^1)^2 - (x'^2)^2}, \quad g_{22} = \frac{a^2 - (x'^1)^2}{a^2 - (x'^1)^2 - (x'^2)^2},$$

$$g_{12} = g_{21} = \frac{x'^1 x'^2}{a^2 - (x'^1)^2 - (x'^2)^2}.$$

Problem 6. For a vector expressed in terms of its covariant components,

$$\vec{v} = v_\alpha \hat{e}^\alpha,$$

show that

$$\partial_\beta \vec{v} = \left(\nabla_\beta v_\alpha\right)\hat{e}^\alpha,$$

where

$$\nabla_\beta v_\alpha = \partial_\beta v_\alpha - \Gamma^\kappa_{\alpha\beta} v^\mu.$$

Problem 7. Suppose the vector, $\vec{v}$, depending on the parameter u on a curve defined by $x^\mu(u)$ is expressed in terms of its covariant components:

$$\vec{v} = v_\mu(u)\hat{e}^\mu(u).$$

Show that the intrinsic derivative of this vector is given by

$$\frac{Dv_\mu}{Du} = \frac{dv_\mu}{du} - \Gamma^\beta_{\mu\kappa} v_\beta \frac{dx^\kappa}{du}.$$

Problem 8. If we change the affine parameter u to u', the coordinates that define the geodesic curve would change from $x^\mu(u)$ to $x^\mu(u')$. Show that in terms of the new affine parameter, u', the geodesic in equation (2.440) becomes

$$\frac{d^2x^\mu}{du'^2} + \Gamma^\mu_{\kappa\beta}\frac{dx^\kappa}{du'}\frac{dx^\beta}{du'} = \left(\frac{\dfrac{d^2u}{du'^2}}{\dfrac{du}{du'}}\right)\frac{dx^\mu}{du'}.$$

Problem 9. Following a similar procedure in example 2.6 show that the four-acceleration of Alien 1 as observed by

(a) Alien 2 in the S' frame is given by

$$a'_x = \frac{a_x}{\gamma_v^3\left(1 - \dfrac{vu_x}{c^2}\right)^3}, \quad a'_y = \frac{1}{\gamma_v^2\left(1 - \dfrac{vu_x}{c^2}\right)^2}a_y + \frac{u_y v}{\gamma_v^2 c^2\left(1 - \dfrac{vu_x}{c^2}\right)^3}a_x,$$

$$a'_z = \frac{1}{\gamma_v^2\left(1 - \dfrac{vu_x}{c^2}\right)^2}a_z + \frac{u_z v}{\gamma_v^2 c^2\left(1 - \dfrac{vu_x}{c^2}\right)^3}a_x.$$

(b) Find these accelerations as observed by the man in the S frame.

Hint: the four-acceleration can be defined in terms of the four-velocity as

$$[a^\alpha] = \frac{du^\alpha}{d\tau} = \gamma_u \frac{du^\alpha}{dt} = \gamma_u \frac{d}{dt}[\gamma_u c, \gamma_u \vec{u}],$$

which gives

$$[a^\alpha] = \gamma_u \frac{d}{dt}[\gamma_u c, \gamma_u \vec{u}] = \gamma_u \left[c \frac{d\gamma_u}{dt}, \frac{d}{dt}(\gamma_u \vec{u}) \right]$$

$$= \gamma_u \left[c \frac{d\gamma_u}{dt}, \vec{u} \frac{d\gamma_u}{dt} + \gamma_u \frac{d\vec{u}}{dt} \right]$$

$$\Rightarrow [a^\alpha] = \gamma_u \left[c \frac{d\gamma_u}{dt}, \vec{u} \frac{d\gamma_u}{dt} + \gamma_u \frac{d\vec{u}}{dt} \right] = \gamma_u \left[c \frac{d\gamma_u}{dt}, \gamma_u \vec{a} + \vec{u} \frac{d\gamma_u}{dt} \right].$$

Problem 10. Derive equation (2.413):

$$\bar{\nu} = \nu \left[1 + \frac{h\nu}{m_0 c^2}(1 - \cos(\theta)) \right]^{-1}.$$

IOP Publishing

Studies in Theoretical Physics, Volume 2
Advanced mathematical methods
Daniel Erenso

Chapter 3

Tensor calculus on manifolds

This chapter introduces tensor calculus on manifolds, offering a thorough exploration of fundamental concepts. Starting with a foundational understanding of tensors
and their ranks, the discussion progresses to unveil the components of tensors,
considering permutations and symmetries within these multidimensional structures.
The chapter explores associated tensors, elucidating the relationships and transformations between different tensor types. Mapping tensors onto tensors is examined in detail, highlighting the crucial role of these mathematical entities in manifold
calculus. Tensors and their behavior under coordinate transformations are analyzed,
providing a comprehensive view of their geometric implications. Tensor equations
and the quotient theorem are explored, shedding light on the algebraic manipu-
lations and relations governing tensors. The chapter discusses covariant derivatives
of tensors, emphasizing their role in understanding the geometric properties of
manifolds. The concept of intrinsic derivatives is introduced, offering a unique
perspective on derivatives within the manifold context. To reinforce and apply the
acquired knowledge, a homework assignment is included.

3.1 Tensors and rank of a tensor

In order to understand what a tensor is and what is its rank is, it is important to have
a better understanding of a vector field, $\vec{v}$, on a manifold. How do we define a vector
field on a manifold? We already know that a vector field at a given point P on a
manifold is defined by the tangent space, T_P, at that point on the manifold. This
tangent space is defined by the tangent vector $\vec{t}$, which can be expressed in terms of
the basis $\hat{e}_\alpha$ or dual basis vectors $\hat{e}^\alpha$.

Inner product

Suppose we want to map the tangent vector $\vec{t}$ onto the vector $\vec{v}$. In other words, we
want to express the vector $\vec{t}$ in terms of whatever basis vectors are used to express the

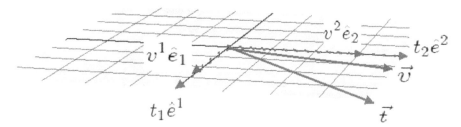

Figure 3.1. Mapping the tangent vector $\vec{t}$ into a vector $\vec{v}$.

vector $\vec{v}$, perhaps with the same basis or dual basis vectors. The projection of vector $\vec{t}$ onto $\vec{v}$ or vice versa is given by the inner product:

$$t(\vec{v}) = \vec{t} \cdot \vec{v}. \tag{3.1}$$

This maps the tangent vector $\vec{t}$ onto the vector field $\vec{v}$. In this case, we say the vector $\vec{t}$ is linearly mapped onto the vector $\vec{v}$ (i.e., $\vec{t} \rightarrow \vec{t}(\vec{v})$) and the vector $\vec{t}(v)$ obtained by mapping $\vec{t}$ onto $\vec{v}$ forms a first-rank tensor. Let us consider a two-dimensional manifold with a plane tangent space where the basis vectors are $\hat{e}_1$, $\hat{e}_2$ and the dual basis vectors $\hat{e}^1$, $\hat{e}^2$ as shown in figure 3.1. The tangent vector, $\vec{t}$, and the vector, $\vec{v}$, at a given point on the manifold can be written as

$$\vec{t} = t_1\hat{e}^1 + t_2\hat{e}^2, \; \vec{v} = v^1\hat{e}_1 + v^2\hat{e}_2,$$

so that

$$\begin{aligned} t^1(\vec{v}) &= \vec{v} \cdot t_1\hat{e}^1 = (v^1\hat{e}_1 + v^2\hat{e}_2) \cdot t_1\hat{e}^1 = v^1 t_1(\hat{e}_1 \cdot \hat{e}^1) + v^2 t_1(\hat{e}_2 \cdot \hat{e}^1) \\ &\Rightarrow t^1(\vec{v}) = v^1 t_1(\hat{e}_1 \cdot \hat{e}^1), \\ t^2(\vec{v}) &= \vec{v} \cdot t_2\hat{e}^2 = (v^1\hat{e}_1 + v^2\hat{e}_2) \cdot t_2\hat{e}^2 = v^1 t_2(\hat{e}_1 \cdot \hat{e}^2) + v^2 t_2(\hat{e}_2 \cdot \hat{e}^2) \\ &\Rightarrow t^2(\vec{v}) = v^2 t_2(\hat{e}_2 \cdot \hat{e}^2). \end{aligned} \tag{3.2}$$

Introducing the column matrices:

$$t^1(\vec{v}) = \vec{v} \cdot t_1\hat{e}^1 = \begin{pmatrix} v^1 t_1(\hat{e}_1 \cdot \hat{e}^1) \\ 0 \end{pmatrix}, \; t^2(\vec{v}) = \vec{v} \cdot t_2\hat{e}^2 = \begin{pmatrix} 0 \\ v^2 t_2(\hat{e}_2 \cdot \hat{e}^2) \end{pmatrix}, \tag{3.3}$$

one can write

$$\vec{t}(\vec{v}) = \begin{pmatrix} v^1 t_1(\hat{e}_1 \cdot \hat{e}^1) \\ 0 \end{pmatrix} + \begin{pmatrix} 0 \\ v^2 t_2(\hat{e}_2 \cdot \hat{e}^2) \end{pmatrix} = \begin{pmatrix} v^1 t_1(\hat{e}_1 \cdot \hat{e}^1) \\ v^2 t_2(\hat{e}_2 \cdot \hat{e}^2) \end{pmatrix}$$

$$\Rightarrow \vec{t}(\vec{v}) = \begin{pmatrix} v^1 t_1(\hat{e}_1 \cdot \hat{e}^1) \\ v^2 t_2(\hat{e}_2 \cdot \hat{e}^2) \end{pmatrix} = \begin{pmatrix} v^1 t_1 \\ v^2 t_2 \end{pmatrix} = \begin{pmatrix} t^1(\vec{v}) \\ t^2(\vec{v}) \end{pmatrix}. \tag{3.4}$$

In a similar manner, we can construct a second-rank tensor from the inner product of a vector, for example the tangent vector $\vec{t}$ that maps two vectors $\vec{u}$ and $\vec{v}$ ($\vec{t}(\vec{u})$ and $\vec{t}(\vec{v})$) to the real numbers given by $t(\vec{u}, \vec{v})$. To this end, consider the same tangent vector $\vec{t}$ that is mapped to the vector, $\vec{u}$,

$$\vec{t}(\vec{u}) = t^1(\vec{u})\hat{e}_1 + t^2(\vec{u})\hat{e}_2, \tag{3.5}$$

and to the vector, $\vec{v}$,

$$\vec{t}(\vec{v}) = t^1(\vec{v})\hat{e}_1 + t^2(\vec{v})\hat{e}_2. \tag{3.6}$$

Using equations (3.5) and (3.6), one can write

$$
\begin{aligned}
t(\vec{u}, \vec{v}) = \vec{t}(\vec{u}) \cdot \vec{t}(\vec{v}) &= (t^1(\vec{u})\hat{e}_1 + t^2(\vec{u})\hat{e}_2) \cdot (t^1(\vec{v})\hat{e}_1 + t^2(\vec{v})\hat{e}_2) \\
&= t^1(\vec{u})t^1(\vec{v})(\hat{e}_1 \cdot \hat{e}_1) + t^1(\vec{u})t^2(\vec{v})(\hat{e}_1 \cdot \hat{e}_2) + t^2(\vec{u})t^1(\vec{v})(\hat{e}_2 \cdot \hat{e}_1) \\
&\quad + t^2(\vec{u})t^2(\vec{v})(\hat{e}_2 \cdot \hat{e}_2).
\end{aligned} \tag{3.7}
$$

Introducing the matrices defined by

$$
\begin{aligned}
t^{11}(\vec{u}, \vec{v}) &= t^1(\vec{u})t^1(\vec{v})(\hat{e}_1 \cdot \hat{e}_1) = \begin{bmatrix} t^1(\vec{u})t^1(\vec{v})(\hat{e}_1 \cdot \hat{e}_1) & 0 \\ 0 & 0 \end{bmatrix}, \\
t^{12}(\vec{u}, \vec{v}) &= t^1(\vec{u})t^2(\vec{v})(\hat{e}_1 \cdot \hat{e}_2) = \begin{bmatrix} 0 & t^1(\vec{u})t^2(\vec{v})(\hat{e}_1 \cdot \hat{e}_2) \\ 0 & 0 \end{bmatrix}, \\
t^{21}(\vec{u}, \vec{v}) &= t^2(\vec{u})t^1(\vec{v})(\hat{e}_2 \cdot \hat{e}_1) = \begin{bmatrix} 0 & 0 \\ t^2(\vec{u})t^1(\vec{v})(\hat{e}_2 \cdot \hat{e}_1) & 0 \end{bmatrix}, \\
t^{22}(\vec{u}, \vec{v}) &= t^2(\vec{u})t^2(\vec{v})(\hat{e}_2 \cdot \hat{e}_2) = \begin{bmatrix} 0 & 0 \\ 0 & t^2(\vec{u})t^2(\vec{v})(\hat{e}_2 \cdot \hat{e}_2) \end{bmatrix},
\end{aligned} \tag{3.8}
$$

one can construct a second-rank tensor:

$$\overset{\leftrightarrow}{t}(\vec{u}, \vec{v}) = \begin{bmatrix} t^1(\vec{u})t^1(\vec{v})(\hat{e}_1 \cdot \hat{e}_1) & t^1(\vec{u})t^2(\vec{v})(\hat{e}_1 \cdot \hat{e}_2) \\ t^2(\vec{u})t^1(\vec{v})(\hat{e}_2 \cdot \hat{e}_1) & t^2(\vec{u})t^2(\vec{v})(\hat{e}_2 \cdot \hat{e}_2) \end{bmatrix} = \begin{bmatrix} t^{11}(\vec{u}, \vec{v}) & t^{12}(\vec{u}, \vec{v}) \\ t^{21}(\vec{u}, \vec{v}) & t^{22}(\vec{u}, \vec{v}) \end{bmatrix}. \tag{3.9}$$

Note that for

$$t^1(\vec{u}) = t^1(\vec{v}) = t^2(\vec{u}) = t^2(\vec{v}) = 1, \tag{3.10}$$

from equation (3.9), one finds

$$\overset{\leftrightarrow}{t}(\vec{u}, \vec{v}) = \begin{bmatrix} \hat{e}_1 \cdot \hat{e}_1 & \hat{e}_1 \cdot \hat{e}_2 \\ \hat{e}_2 \cdot \hat{e}_1 & \hat{e}_2 \cdot \hat{e}_2 \end{bmatrix} = [g_{\alpha\beta}], \tag{3.11}$$

which is the covariant component of the second-rank metric tensor.

Outer product
The inner product of two vectors is a scalar, which is a zero-rank tensor. However, unlike the inner product, the outer product, which is also referred to as a tensor product, results in a second-rank tensor. For two vectors, $\vec{u}(\vec{p})$ and $\vec{v}(\vec{q})$, the outer product, denoted by $\vec{u} \otimes \vec{v} (\vec{p}, \vec{q})$, is given by

$$(\vec{u} \otimes \vec{v})(\vec{p}, \vec{q}) = \vec{u}(\vec{p})\vec{v}(\vec{q}). \tag{3.12}$$

Suppose $(\vec{p}, \vec{q})$ are expressed in terms of the dual basis vectors $(\hat{e}^\alpha, \hat{e}^\beta)$:

$$\vec{u}(\hat{e}_\alpha) = u_\alpha \hat{e}^\alpha, \; \vec{v}(\hat{e}_\beta) = u_\beta \hat{e}^\beta, \tag{3.13}$$

one can write

$$
\begin{aligned}
(\vec{u} \otimes \vec{v})(\hat{e}_\alpha, \hat{e}_\beta) &= \vec{u}(\hat{e}_\alpha) \otimes \vec{v}(\hat{e}_\beta) = u_\alpha v_\beta (\hat{e}^\alpha \otimes \hat{e}^\beta) \\
\Rightarrow (\vec{u} \otimes \vec{v})(\hat{e}_\alpha, \hat{e}_\beta) &= [u_\alpha v_\beta].
\end{aligned} \tag{3.14}
$$

For example, for vectors with two components, we have

$$(\vec{u} \otimes \vec{v})(\hat{e}_\alpha, \hat{e}_\beta) = u_1 v_1 (\hat{e}^1 \otimes \hat{e}^1) + u_1 v_2 (\hat{e}^1 \otimes \hat{e}^2)$$
$$+ u_2 v_1 (\hat{e}^2 \otimes \hat{e}^1) + u_2 v_2 (\hat{e}^2 \otimes \hat{e}^2). \tag{3.15}$$

Introducing the matrices

$$\hat{e}^1 \otimes \hat{e}^1 = \begin{pmatrix} \hat{e}^1 \otimes \hat{e}^1 & 0 \\ 0 & 0 \end{pmatrix}, \quad \hat{e}^1 \otimes \hat{e}^2 = \begin{pmatrix} 0 & \hat{e}^1 \otimes \hat{e}^2 \\ 0 & 0 \end{pmatrix},$$
$$\hat{e}^2 \otimes \hat{e}^1 = \begin{pmatrix} 0 & 0 \\ \hat{e}^2 \otimes \hat{e}^1 & 0 \end{pmatrix}, \quad \hat{e}^2 \otimes \hat{e}^2 = \begin{pmatrix} 0 & 0 \\ 0 & \hat{e}^2 \otimes \hat{e}^2 \end{pmatrix}, \tag{3.16}$$

one can write

$$(\vec{u} \otimes \vec{v})(\hat{e}_\alpha, \hat{e}_\beta) = u_1 v_1 \begin{pmatrix} \hat{e}^1 \otimes \hat{e}^1 & 0 \\ 0 & 0 \end{pmatrix} + u_1 v_2 \begin{pmatrix} 0 & \hat{e}^1 \otimes \hat{e}^2 \\ 0 & 0 \end{pmatrix}$$
$$+ u_2 v_1 \begin{pmatrix} 0 & 0 \\ \hat{e}^2 \otimes \hat{e}^1 & 0 \end{pmatrix} + u_2 v_2 \begin{pmatrix} 0 & 0 \\ 0 & \hat{e}^2 \otimes \hat{e}^2 \end{pmatrix} \tag{3.17}$$
$$\Rightarrow (\vec{u} \otimes \vec{v})(\hat{e}_\alpha, \hat{e}_\beta) = \begin{pmatrix} u_1 v_1 (\hat{e}^1 \otimes \hat{e}^1) & u_1 v_2 (\hat{e}^1 \otimes \hat{e}^2) \\ u_2 v_1 (\hat{e}^2 \otimes \hat{e}^1) & u_2 v_2 (\hat{e}^2 \otimes \hat{e}^2) \end{pmatrix} = [u_\alpha u_\beta] = [t_{\alpha\beta}].$$

The tensor product of these two vectors, each with two components, can be visualized as shown in figure 3.2. Let the components for the vector $\vec{u}$ be denoted by the green couple of a male and a female, and the components of the vector $\vec{v}$ by the blue couple. In these rank-1 tensors each person stays in their 'room' (the red boxes). However, the tensor product of these two vectors, as can be seen in figure 3.2, creates a second-rank tensor that results in the existence of two persons from each in a room. We can also take the out product of tensors with different rank. Now let us consider a first-rank tensor $\vec{s}(\vec{r})$ and a second-rank tensor $\overleftrightarrow{t}(\vec{p}, \vec{q})$. The outer product of these two tensors gives a rank-3 tensor, given by

$$\overleftrightarrow{t} \otimes \vec{s}(\vec{p}, \vec{q}, \vec{r}) = t(\vec{p}, \vec{q})s(\vec{r}). \tag{3.18}$$

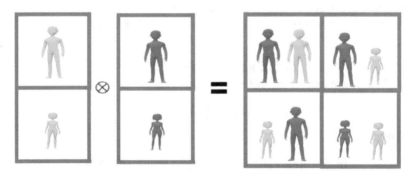

Figure 3.2. The second-rank tensor resulting from a tensor product of two two-dimensional vectors.

Again if the vectors $(\vec{p}, \vec{q}, \vec{r})$ are expressed in terms of the dual basis vectors $(\hat{e}^\alpha, \hat{e}^\beta, \hat{e}^\kappa)$, one can write

$$\overset{\leftrightarrow}{t} \otimes \vec{s}(\hat{e}^\alpha, \hat{e}^\beta, \hat{e}^\kappa) = t(\hat{e}^\alpha, \hat{e}^\beta)s(\hat{e}^\kappa) = t_{\alpha\beta}s_\kappa = h_{\alpha\beta\kappa}. \tag{3.19}$$

Similarly, for the case where the basis and dual basis vectors are mixed, we can express the components as

$$\overset{\leftrightarrow}{t} \otimes \vec{s}(\hat{e}_\alpha, \hat{e}_\beta, \hat{e}_\kappa) = t(\hat{e}_\alpha, \hat{e}_\beta)s(\hat{e}_\kappa) = t^{\alpha\beta}s^\kappa = h^{\alpha\beta\kappa},$$

$$\overset{\leftrightarrow}{t} \otimes \vec{s}(\hat{e}_\alpha, \hat{e}^\beta, \hat{e}^\kappa) = t(\hat{e}_\alpha, \hat{e}^\beta)s(\hat{e}^\kappa) = t^\alpha_\beta s_\kappa = h^\alpha_{\beta\kappa},$$

$$\overset{\leftrightarrow}{t} \otimes \vec{s}(\hat{e}_\alpha, \hat{e}_\beta, \hat{e}^\kappa) = t(\hat{e}_\alpha, \hat{e}_\beta)s(\hat{e}^\kappa) = t^{\alpha\beta}s_\kappa = h^{\alpha\beta}_\kappa, \tag{3.20}$$

$$\overset{\leftrightarrow}{t} \otimes \vec{s}(\hat{e}^\alpha, \hat{e}^\beta, \hat{e}_\kappa) = t(\hat{e}^\alpha, \hat{e}^\beta)s(\hat{e}_\kappa) = t_{\alpha\beta}s^\kappa = h^\kappa_{\alpha\beta}.$$

To better understand the tensor product of a second-rank tensor (which is a tensor product of two first-rank tensors) and a rank-1 tensor, let the second-rank tensor, $\overset{\leftrightarrow}{t}$, be the tensor we created from green and blue couples and the rank-1 tensor, $\vec{s}$, be the yellow couple. As we can see in figure 3.3, this tensor product creates eight rooms (i.e., components) where any three persons from each can coexist in a room.

Tensor	Rank
$\vec{t}(\vec{u})$	1
$\overset{\leftrightarrow}{t}(\vec{u}, \vec{v})$	2
$t(\vec{u}, \vec{v}, \vec{w})$	3
$t(\vec{u}, \vec{v}, \vec{w}, \vec{x})$	4
$t(\vec{u}, \vec{v}, \vec{w}, \vec{x}, \vec{y})$	5

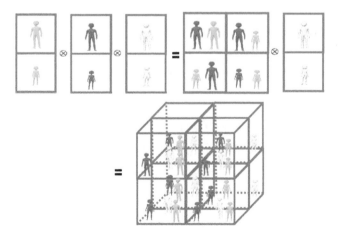

Figure 3.3. A third-rank tensor resulting from a tensor product of three two-dimensional vectors (rank-1 tensors) or a tensor product of rank-2 and a rank-1 tensors.

Generally, a tensor is defined by the precise set of operations applied to a set of vectors, and the number of vectors in the set determines the rank of the tensor. If there are N number of vectors in a set, the tensor is an Nth rank tensor.

From this table we can easily see that a scalar field can be classified as a *zero-rank* tensor field since it does not depend on a vector field. A tensor is a linear map of a set of vectors and therefore any ranked tensor is linear. This means, for example, for a first-rank tensor we must have

$$\vec{t}(\alpha\vec{u} + \beta\vec{v}) = \vec{t}(\alpha\vec{u}) + \vec{t}(\beta\vec{v}) = \alpha\vec{t}(\vec{u}) + \beta\vec{t}(\vec{v}). \tag{3.21}$$

Addition and subtraction

We can add or subtract tensors of the same rank. Suppose $\overleftrightarrow{r}(\hat{e}_\alpha, \hat{e}_\beta)$ and $\overleftrightarrow{s}(\hat{e}_\alpha, \hat{e}_\beta)$ are rank-2 tensors. Then by adding or subtracting these two tensors, one can find a rank-2 tensor $\overleftrightarrow{t}(\hat{e}_\alpha, \hat{e}_\beta)$:

$$\overleftrightarrow{t}(\hat{e}_\alpha, \hat{e}_\beta) = \overleftrightarrow{r}(\hat{e}_\alpha, \hat{e}_\beta) \pm \overleftrightarrow{s}(\hat{e}_\alpha, \hat{e}_\beta)$$
$$\Rightarrow t_{\alpha\beta} = r_{\alpha\beta} \pm s_{\alpha\beta}. \tag{3.22}$$

Multiplying a tensor with a scalar

Multiplying a tensor of any rank with a scalar results in multiplying all of its components. For example, multiplying $\overleftrightarrow{r}(\hat{e}_\alpha, \hat{e}_\beta)$ and $\overleftrightarrow{s}(\hat{e}_\alpha, \hat{e}_\beta)$ with α results in

$$\alpha\overleftrightarrow{t}(\hat{e}_\alpha, \hat{e}_\beta) = \alpha\overleftrightarrow{r}(\hat{e}_\alpha, \hat{e}_\beta) \pm \alpha\overleftrightarrow{s}(\hat{e}_\alpha, \hat{e}_\beta)$$
$$\Rightarrow \alpha t_{\alpha\beta} = \alpha r_{\alpha\beta} \pm \alpha s_{\alpha\beta}. \tag{3.23}$$

3.2 Components of a tensor

We recall that the tangent space is defined in terms of the basis $(\hat{e}_\alpha)$ or dual basis vectors $(\hat{e}^\beta)$. When vectors are expressed in terms of basis or dual basis vectors, we can determine the components of a tensor in different forms. But first let us consider when the vectors $\vec{v}$ and $\vec{u}$ are the basis or the dual basis vectors. In this case, we have as follows.

(a) First-rank tensor:

$$t(\hat{e}_\alpha) = \vec{t} \cdot \hat{e}_\alpha = t_\alpha,$$
$$t(\hat{e}^\beta) = \vec{t} \cdot \hat{e}^\beta = t^\beta. \tag{3.24}$$

(b) Second-rank tensor:

$$t(\hat{e}^\alpha, \hat{e}^\beta) = \vec{t}(\hat{e}^\alpha) \cdot \vec{t}(\hat{e}^\beta) = t^{\alpha\beta},$$
$$t(\hat{e}_\alpha, \hat{e}_\beta) = \vec{t}(\hat{e}_\alpha) \cdot \vec{t}(\hat{e}_\beta) = t_{\alpha\beta},$$
$$t(\hat{e}_\alpha, \hat{e}^\beta) = \vec{t}(\hat{e}_\alpha) \cdot \vec{t}(\hat{e}^\beta) = t_\alpha^{\ \beta},$$
$$t(\hat{e}^\alpha, \hat{e}_\beta) = \vec{t}(\hat{e}^\alpha) \cdot \vec{t}(\hat{e}_\beta) = t_\beta^{\ \alpha}. \tag{3.25}$$

Now, we can apply the linearity of tensors to determine the components of a tensor for the general case where we have two vectors $\vec{v}$ and $\vec{u}$ expressed in terms of the covariant or contravariant components as

$$\vec{u} = u_\alpha \hat{e}^\alpha = u^\alpha \hat{e}_\alpha, \quad \vec{v} = v_\beta \hat{e}^\beta = v^\beta \hat{e}_\beta.$$

(a) Rank-1 tensor:

$$t(\vec{u}) = t(u_\alpha \hat{e}^\alpha) = t(\hat{e}^\alpha)u_\alpha = t^\alpha u_\alpha,$$
$$t(\vec{u}) = t(u^\alpha \hat{e}_\alpha) = u^\alpha t(\hat{e}_\alpha) = t_\alpha u^\alpha.$$

(3.26)

(b) Rank-2 tensor:

$$t(\vec{u}, \vec{v}) = t\left(u^\alpha \hat{e}_\alpha, v_\beta \hat{e}^\beta\right) = t(\hat{e}_\alpha, \hat{e}^\beta)u^\alpha v_\beta = u^\alpha t_\alpha^\beta v_\beta,$$
$$t(\vec{u}, \vec{v}) = t\left(u_\alpha \hat{e}^\alpha, v^\beta \hat{e}_\beta\right) = t(\hat{e}^\alpha, \hat{e}_\beta)u_\alpha v^\beta = v^\beta t_\beta^\alpha u_\alpha,$$
$$t(\vec{u}, \vec{v}) = t\left(u^\alpha \hat{e}_\alpha, v^\beta \hat{e}_\beta\right) = t(\hat{e}_\alpha, \hat{e}_\beta)u^\alpha v^\beta = u^\alpha t_{\alpha\beta} v^\beta.$$

(3.27)

Example 3.1. Let us reconsider the two-dimensional sphere in the three-dimensional Euclidian manifold shown in figure 3.4. For a two-dimensional sphere we recall that

$$\hat{e}_1(x^1, x^2) = \cos(x^2)\hat{x} + \sin(x^2)\hat{y} - \frac{x^1}{\sqrt{a^2 - (x^1)^2}}\hat{z}.$$
$$\hat{e}_2(x^1, x^2) = -x^1 \sin(x^2)\hat{x} + x^1 \cos(x^2)\hat{y}.$$

(3.28)

Find the metric tensor components $g_{\alpha\beta}$, g_β^α, g_α^β, and $g^{\alpha\beta}$.

Solution: One can define the tangent vector shown in figure 3.4,

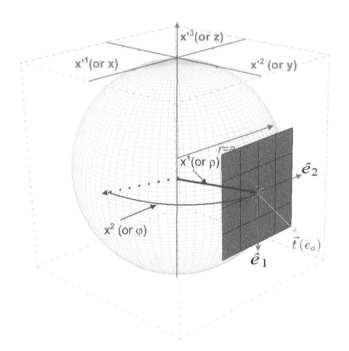

Figure 3.4. A two-dimensional sphere in three-dimensional Euclidean space.

$$\vec{t}(\hat{e}_a) = \hat{e}_1 + \hat{e}_2 = \hat{e}^1 + \hat{e}^2, \tag{3.29}$$

and write the covariant components for the metric tensor $g_{\alpha\beta}$ as

$$\overleftrightarrow{t}(\hat{e}_a, \hat{e}_\beta) = \begin{bmatrix} \hat{e}_1 \cdot \hat{e}_1 & \hat{e}_1 \cdot \hat{e}_2 \\ \hat{e}_2 \cdot \hat{e}_1 & \hat{e}_2 \cdot \hat{e}_2 \end{bmatrix} = \begin{bmatrix} g_{11} & g_{12} \\ g_{21} & g_{22} \end{bmatrix} = [g_{\alpha\beta}]. \tag{3.30}$$

Using equation (3.28), we find

$$g_{11} = \hat{e}_1 \cdot \hat{e}_1 = \frac{a^2}{a^2 - (x^{'1})^2}, \; g_{22} = \hat{e}_2 \cdot \hat{e}_2 = (x^1)^2,$$

$$g_{12} = g_{21} = \hat{e}_1 \cdot \hat{e}_2 = 0. \tag{3.31}$$

For the contravariant component

$$\overleftrightarrow{t}(\hat{e}^\alpha, \hat{e}^\beta) = \begin{bmatrix} \hat{e}^1 \cdot \hat{e}^1 & \hat{e}^1 \cdot \hat{e}^2 \\ \hat{e}^2 \cdot \hat{e}^1 & \hat{e}^2 \cdot \hat{e}^2 \end{bmatrix} = \begin{bmatrix} g^{12} & g^{12} \\ g^{21} & g^{22} \end{bmatrix}. \tag{3.32}$$

Noting that

$$\hat{e}^\alpha = g^{\alpha\kappa}\hat{e}_\kappa, \; \hat{e}^\beta = g^{\beta\mu}\hat{e}_\mu, \tag{3.33}$$

we have

$$g^{\alpha\beta} = \hat{e}^\alpha \cdot \hat{e}^\beta = g^{\alpha\kappa}\hat{e}_\kappa \cdot g^{\beta\mu}\hat{e}_\mu = g^{\alpha\kappa}g^{\beta\mu}\hat{e}_\kappa \cdot \hat{e}_\mu = g^{\alpha\kappa}g^{\beta\mu}g_{\kappa\mu}$$

$$\Rightarrow g^{\alpha\beta} = g^{\alpha\kappa}g^{\beta\mu}g_{\kappa\mu} = g^{\alpha\kappa}g_{\kappa\mu}g^{\beta\mu}, \tag{3.34}$$

or, in a matrix form,

$$[g^{\alpha\beta}] = [g^{\alpha\kappa}][g_{\kappa\mu}][g^{\beta\mu}]$$

$$\begin{bmatrix} g^{12} & g^{12} \\ g^{21} & g^{22} \end{bmatrix} = \begin{bmatrix} g^{12} & g^{12} \\ g^{21} & g^{22} \end{bmatrix} \begin{bmatrix} g_{11} & g_{12} \\ g_{21} & g_{22} \end{bmatrix} \begin{bmatrix} g^{12} & g^{12} \\ g^{21} & g^{22} \end{bmatrix}. \tag{3.35}$$

This equation holds true if and only if

$$\begin{bmatrix} g_{11} & g_{12} \\ g_{21} & g_{22} \end{bmatrix} \begin{bmatrix} g^{12} & g^{12} \\ g^{21} & g^{22} \end{bmatrix} = \begin{bmatrix} 1 & 0 \\ 0 & 1 \end{bmatrix} = I.$$

There follows that

$$\begin{bmatrix} g^{12} & g^{12} \\ g^{21} & g^{22} \end{bmatrix} = \begin{bmatrix} g_{11} & g_{12} \\ g_{21} & g_{22} \end{bmatrix}^{-1}.$$

We recall for a 2×2 matrix,

$$G = \begin{bmatrix} g_{11} & g_{12} \\ g_{21} & g_{22} \end{bmatrix}, \tag{3.36}$$

the inverse matrix is given by

$$G^{-1} = \frac{adj(G)}{|G|} = \frac{\begin{pmatrix} g_{22} & -g_{12} \\ -g_{21} & g_{11} \end{pmatrix}}{g_{22}g_{11} - g_{12}g_{21}}, \tag{3.37}$$

so that using the results in equation (3.31) one finds

$$\begin{bmatrix} g^{12} & g^{12} \\ g^{21} & g^{22} \end{bmatrix} = \begin{bmatrix} g_{11} & g_{12} \\ g_{21} & g_{22} \end{bmatrix}^{-1} = \frac{\begin{pmatrix} g_{22} & 0 \\ 0 & g_{11} \end{pmatrix}}{g_{22}g_{11}}$$

$$\Rightarrow \begin{bmatrix} g^{12} & g^{12} \\ g^{21} & g^{22} \end{bmatrix} = \begin{pmatrix} \dfrac{1}{g_{11}} & 0 \\ 0 & \dfrac{1}{g_{22}} \end{pmatrix}. \tag{3.38}$$

Using the dual basis vector, we can also express this second-rank tensor as

$$\overset{\leftrightarrow}{t}(\hat{e}_\alpha, \hat{e}^\beta) = \overset{\leftrightarrow}{t}(\hat{e}^\alpha, \hat{e}_\beta) = \begin{bmatrix} \hat{e}_1 \cdot \hat{e}^1 & \hat{e}_1 \cdot \hat{e}^2 \\ \hat{e}_2 \cdot \hat{e}^1 & \hat{e}_2 \cdot \hat{e}^2 \end{bmatrix} = \begin{bmatrix} g_1^1 & g_1^2 \\ g_2^1 & g_2^2 \end{bmatrix} \tag{3.39}$$

Recalling that

$$\hat{e}^\alpha = g^{\alpha\kappa}\hat{e}_\kappa,$$

one can write

$$g_\beta^{\,\alpha} = \hat{e}^\alpha \cdot \hat{e}_\beta = g^{\alpha\kappa}\hat{e}_\kappa \cdot \hat{e}_\beta = g^{\alpha\kappa}g_{\kappa\beta}$$

$$\Rightarrow [g_\beta^{\,\alpha}] = \begin{bmatrix} g_1^1 & g_1^2 \\ g_2^1 & g_2^2 \end{bmatrix} = \begin{bmatrix} g^{12} & g^{12} \\ g^{21} & g^{22} \end{bmatrix}\begin{bmatrix} g_{11} & g_{12} \\ g_{21} & g_{22} \end{bmatrix}. \tag{3.40}$$

Since we already have shown that

$$\begin{bmatrix} g^{12} & g^{12} \\ g^{21} & g^{22} \end{bmatrix} = \begin{bmatrix} g_{11} & g_{12} \\ g_{21} & g_{22} \end{bmatrix}^{-1} = \begin{bmatrix} \dfrac{1}{g_{11}} & 0 \\ 0 & \dfrac{1}{g_{22}} \end{bmatrix},$$

one can easily find

$$\overset{\leftrightarrow}{t}(\hat{e}_\alpha, \hat{e}^\beta) = \overset{\leftrightarrow}{t}(\hat{e}^\alpha, \hat{e}_\beta) = \begin{bmatrix} \hat{e}_1 \cdot \hat{e}^1 & \hat{e}_1 \cdot \hat{e}^2 \\ \hat{e}_2 \cdot \hat{e}^1 & \hat{e}_2 \cdot \hat{e}^2 \end{bmatrix} = \begin{bmatrix} g_1^1 & g_1^2 \\ g_2^1 & g_2^2 \end{bmatrix}$$

$$= \begin{bmatrix} \dfrac{1}{g_{11}} & 0 \\ 0 & \dfrac{1}{g_{22}} \end{bmatrix} \begin{bmatrix} g_{11} & 0 \\ 0 & g_{22} \end{bmatrix} \tag{3.41}$$

$$\Rightarrow \begin{bmatrix} g_\beta^\alpha \end{bmatrix} = \begin{bmatrix} g_\alpha^\beta \end{bmatrix} \begin{bmatrix} g_1^1 & g_1^2 \\ g_2^1 & g_2^2 \end{bmatrix} = \begin{bmatrix} 1 & 0 \\ 0 & 1 \end{bmatrix}.$$

Example 3.2. Consider two vectors in the tangent space for a point in a three-dimensional sphere embedded in a four-dimensional manifold given by

$$\vec{p} = p^\alpha \hat{e}_\alpha, \; \vec{q} = q^\beta \hat{e}_\beta,$$

where α, $\beta = 1$, 2, 3, and $\hat{e}_1$, $\hat{e}_2$, and $\hat{e}_3$ (or $\hat{e}_r$, $\hat{e}_\theta$, and $\hat{e}_\varphi$). Find the components of the second-rank tensor for

$$u \otimes v(\vec{p}, \vec{q}) = u(\vec{p})v(\vec{q}). \tag{3.42}$$

Solution: The components of this second-rank tensor are given by

$$u \otimes v(\vec{p}, \vec{q}) = p^\alpha q^\beta \hat{e}_\alpha \hat{e}_\beta$$
$$= p^1 q^1 \hat{e}_1 \hat{e}_1 + p^1 q^2 \hat{e}_1 \hat{e}_2 + p^1 q^3 \hat{e}_1 \hat{e}_3$$
$$+ p^2 q^1 \hat{e}_2 \hat{e}_1 + p^2 q^2 \hat{e}_2 \hat{e}_2 + p^3 q^3 \hat{e}_3 \hat{e}_3$$
$$+ p^3 q^1 \hat{e}_3 \hat{e}_1 + p^3 q^2 \hat{e}_3 \hat{e}_2 + p^3 q^3 \hat{e}_3 \hat{e}_3,$$

and using a matrix this can be expressed as

$$u \otimes v(\vec{p}, \vec{q}) = \begin{bmatrix} p^1 q^1 & p^1 q^2 & p^1 q^3 \\ p^2 q^1 & p^2 q^2 & p^3 q^3 \\ p^3 q^1 & p^3 q^2 & p^3 q^3 \end{bmatrix}.$$

Example 3.3. In electrodynamics the force density (force per unit volume), $\vec{f}$, is given by

$$\vec{f} = \nabla \cdot \overset{\leftrightarrow}{T} - \epsilon_0 \mu_0 \frac{\partial \vec{S}}{\partial t}. \tag{3.43}$$

In equation (3.43), $\vec{S}$ is known as the Poynting vector (a rank-1 tensor), which is proportional to the cross product of the electric $\vec{E}$ and magnetic $\vec{B}$ field vectors (rank-1 tensors):

$$\vec{S} = \frac{1}{\mu_0}\left(\vec{E} \times \vec{B}\right), \tag{3.44}$$

where μ_0 is the magnetic permeability of a free space. The quantity $\overset{\leftrightarrow}{T}$ (rank-2 tensor) is called *Maxwell's stress tensor*, and is given by

$$T_{\alpha\beta} = \epsilon_0\left(E_\alpha E_\beta - \frac{1}{2}\delta_{\alpha\beta}E^2\right) + \frac{1}{\mu_0}\left(B_\alpha B_\beta - \frac{1}{2}\delta_{\alpha\beta}B^2\right), \tag{3.45}$$

where E_α and B_α are the components of the electric and magnetic field vectors, respectively, and ϵ_0 is the electrical permittivity of a free space. If we set $\vec{B} = 0$, which give $\vec{S} = 0$, the Maxwell's stress tensor in equation (3.45) reduces to

$$T_{\alpha\beta} = \epsilon_0\left(E_\alpha E_\beta - \frac{1}{2}\delta_{\alpha\beta}E^2\right). \tag{3.46}$$

Using the rank-1 electric field tensor,

$$\vec{E} = E_x\hat{x} + E_y\hat{y} + E_z\hat{z}, \tag{3.47}$$

and the matrices defined by

$$\hat{x} \otimes \hat{x} = \begin{pmatrix} \hat{x} \otimes \hat{x} & 0 & 0 \\ 0 & 0 & 0 \\ 0 & 0 & 0 \end{pmatrix}, \quad \hat{x} \otimes \hat{y} = \begin{pmatrix} 0 & \hat{x} \otimes \hat{y} & 0 \\ 0 & 0 & 0 \\ 0 & 0 & 0 \end{pmatrix},$$

$$\hat{x} \otimes \hat{z} = \begin{pmatrix} 0 & 0 & \hat{x} \otimes \hat{z} \\ 0 & 0 & 0 \\ 0 & 0 & 0 \end{pmatrix},$$

$$\hat{y} \otimes \hat{x} = \begin{pmatrix} 0 & 0 & 0 \\ \hat{y} \otimes \hat{x} & 0 & 0 \\ 0 & 0 & 0 \end{pmatrix}, \quad \hat{y} \otimes \hat{y} = \begin{pmatrix} 0 & 0 & 0 \\ 0 & \hat{y} \otimes \hat{y} & 0 \\ 0 & 0 & 0 \end{pmatrix},$$

$$\hat{y} \otimes \hat{z} = \begin{pmatrix} 0 & 0 & 0 \\ 0 & 0 & \hat{y} \otimes \hat{z} \\ 0 & 0 & 0 \end{pmatrix}, \tag{3.48}$$

$$\hat{z} \otimes \hat{x} = \begin{pmatrix} 0 & 0 & 0 \\ 0 & 0 & 0 \\ \hat{z} \otimes \hat{x} & 0 & 0 \end{pmatrix}, \quad \hat{z} \otimes \hat{y} = \begin{pmatrix} 0 & 0 & 0 \\ 0 & 0 & 0 \\ 0 & \hat{z} \otimes \hat{y} & 0 \end{pmatrix},$$

$$\hat{z} \otimes \hat{z} = \begin{pmatrix} 0 & 0 & 0 \\ 0 & 0 & 0 \\ 0 & 0 & \hat{z} \otimes \hat{z} \end{pmatrix}.$$

(a) Show that Maxwell's stress tensor in equation (3.46) is a rank-2 symmetric tensor given by

$$\overleftrightarrow{T} = \begin{pmatrix} \epsilon_0\left(E_x^2 - \dfrac{1}{2}E^2\right) & \epsilon_0 E_x E_y & \epsilon_0 E_x E_z \\[2mm] \epsilon_0 E_y E_x & \epsilon_0\left(E_y^2 - \dfrac{1}{2}E^2\right) & \epsilon_0 E_y E_z \\[2mm] \epsilon_0 E_z E_x & \epsilon_0 E_z E_y & \epsilon_0\left(E_z^2 - \dfrac{1}{2}E^2\right) \end{pmatrix}. \tag{3.49}$$

(b) If we represent $\hat{x} \to \hat{e}_1$, $\hat{y} \to \hat{e}_2$, and $\hat{z} \to \hat{e}_3$, one can write

$$\vec{E} = E^\alpha \hat{e}_\alpha, \tag{3.50}$$

and

$$\overleftrightarrow{T} = \begin{pmatrix} \epsilon_0\left(E^1 E^1 - \dfrac{1}{2}E^2\right) & \epsilon_0 E^1 E^2 & \epsilon_0 E^1 E^3 \\[2mm] \epsilon_0 E^2 E^1 & \epsilon_0\left(E^2 E^2 - \dfrac{1}{2}E^2\right) & \epsilon_0 E^2 E^3 \\[2mm] \epsilon_0 E^3 E^1 & \epsilon_0 E^3 E^2 & \epsilon_0\left(E^3 E^3 - \dfrac{1}{2}E^2\right) \end{pmatrix}, \tag{3.51}$$

where we used

$$[E^\alpha] = [E^1,\, E^2,\, E^3] = \left[E_x,\, E_y,\, E_z\right].$$

Furthermore, recalling that the divergence for a vector is given by

$$\nabla \cdot \vec{v} = \nabla_\alpha v^\alpha = \partial_\alpha v^\alpha + \Gamma^\alpha_{\beta\alpha} v^\beta, \tag{3.52}$$

one can write the divergence for a rank-2 tensor as

$$\nabla \cdot \overleftrightarrow{T} = \nabla_\alpha T^{\alpha\beta} = \partial_\alpha T^{\alpha\beta} + \Gamma^\alpha_{\mu\alpha} T^{\mu\beta}. \tag{3.53}$$

Find the metric tensor $g_{\alpha\beta}$ and show that $\Gamma^\alpha_{\mu\alpha} = 0$.

(c) Find the force

$$\vec{f} = \nabla \cdot \overleftrightarrow{T} = \nabla_\alpha T^{\alpha\beta} = \partial_\alpha T^{\alpha\beta}. \tag{3.54}$$

Solution:

(a) Noting that

$$\epsilon_0 E_\alpha E_\beta = \epsilon_0 \left(E_x \hat{x} + E_y \hat{y} + E_z \hat{z}\right) \otimes \left(E_x \hat{x} + E_y \hat{y} + E_z \hat{z}\right)$$

$$= \epsilon_0 \{E_x^2 \hat{x} \otimes \hat{x} + E_x E_y \hat{x} \otimes \hat{y} + E_x E_z \hat{x} \otimes \hat{z}$$

$$+ E_y E_x \hat{y} \otimes \hat{x} + E_y^2 \hat{y} \otimes \hat{y} + E_y E_z \hat{y} \otimes \hat{z} \qquad (3.55)$$

$$+ E_z E_x \hat{z} \otimes \hat{x} + E_z E_y \hat{z} \otimes \hat{y} + E_z^2 \hat{z} \otimes \hat{z}\},$$

and substituting

$$\hat{x} \otimes \hat{x} = \begin{pmatrix} \hat{x} \otimes \hat{x} & 0 & 0 \\ 0 & 0 & 0 \\ 0 & 0 & 0 \end{pmatrix}, \quad \hat{x} \otimes \hat{y} = \begin{pmatrix} 0 & \hat{x} \otimes \hat{y} & 0 \\ 0 & 0 & 0 \\ 0 & 0 & 0 \end{pmatrix},$$

$$\hat{x} \otimes \hat{z} = \begin{pmatrix} 0 & 0 & \hat{x} \otimes \hat{z} \\ 0 & 0 & 0 \\ 0 & 0 & 0 \end{pmatrix},$$

$$\hat{y} \otimes \hat{x} = \begin{pmatrix} 0 & 0 & 0 \\ \hat{y} \otimes \hat{x} & 0 & 0 \\ 0 & 0 & 0 \end{pmatrix}, \quad \hat{y} \otimes \hat{y} = \begin{pmatrix} 0 & 0 & 0 \\ 0 & \hat{y} \otimes \hat{y} & 0 \\ 0 & 0 & 0 \end{pmatrix},$$

$$\hat{y} \otimes \hat{z} = \begin{pmatrix} 0 & 0 & 0 \\ 0 & 0 & \hat{y} \otimes \hat{z} \\ 0 & 0 & 0 \end{pmatrix}, \qquad (3.56)$$

$$\hat{z} \otimes \hat{x} = \begin{pmatrix} 0 & 0 & 0 \\ 0 & 0 & 0 \\ \hat{z} \otimes \hat{x} & 0 & 0 \end{pmatrix}, \quad \hat{z} \otimes \hat{y} = \begin{pmatrix} 0 & 0 & 0 \\ 0 & 0 & 0 \\ 0 & \hat{z} \otimes \hat{y} & 0 \end{pmatrix},$$

$$\hat{z} \otimes \hat{z} = \begin{pmatrix} 0 & 0 & 0 \\ 0 & 0 & 0 \\ 0 & 0 & \hat{z} \otimes \hat{z} \end{pmatrix},$$

or, in much simpler form,

$$\overset{\leftrightarrow}{T} = \begin{pmatrix} \epsilon_0 E_x^2 & \epsilon_0 E_x E_y & \epsilon_0 E_x E_z \\ \epsilon_0 E_y E_x & \epsilon_0 E_y^2 & \epsilon_0 E_y E_z \\ \epsilon_0 E_z E_x & \epsilon_0 E_z E_y & \epsilon_0 E_z^2 \end{pmatrix}, \qquad (3.57)$$

where we omitted $\hat{x} \otimes \hat{x}, \hat{y} \otimes \hat{y}, \hat{z} \otimes \hat{z}, \hat{x} \otimes \hat{z}, \hat{z} \otimes \hat{x}, \hat{x} \otimes \hat{y}, \hat{y} \otimes \hat{x}, \hat{y} \otimes \hat{z}$, and $\hat{z} \otimes \hat{y}$. For the second term in equation (3.46), one can write

$$\frac{1}{2}\delta_{\alpha\beta}\epsilon_0 E^2 = \frac{1}{2}\epsilon_0 E^2 \begin{bmatrix} \hat{x} \otimes \hat{x} & 0 & 0 \\ 0 & \hat{y} \otimes \hat{y} & 0 \\ 0 & 0 & \hat{z} \otimes \hat{z} \end{bmatrix},$$

and we find

$$
\overset{\leftrightarrow}{T} = \begin{pmatrix}
\epsilon_0\left(E_x^2 - \dfrac{1}{2}E^2\right) & \epsilon_0 E_x E_y & \epsilon_0 E_x E_z \\[2ex]
\epsilon_0 E_y E_x & \epsilon_0\left(E_y^2 - \dfrac{1}{2}E^2\right) & \epsilon_0 E_y E_z \\[2ex]
\epsilon_0 E_z E_x & \epsilon_0 E_z E_y & \epsilon_0\left(E_z^2 - \dfrac{1}{2}E^2\right)
\end{pmatrix}. \tag{3.58}
$$

(b) Using $\hat{x} \to \hat{e}_1$, $\hat{y} \to \hat{e}_2$, and $\hat{z} \to \hat{e}_3$, one can write $g_{\alpha\beta}$:

$$
[g_{\alpha\beta}] = \begin{bmatrix}
\hat{e}_1 \cdot \hat{e}_1 & \hat{e}_1 \cdot \hat{e}_2 & \hat{e}_1 \cdot \hat{e}_3 \\
\hat{e}_2 \cdot \hat{e}_1 & \hat{e}_2 \cdot \hat{e}_2 & \hat{e}_2 \cdot \hat{e}_3 \\
\hat{e}_3 \cdot \hat{e}_1 & \hat{e}_3 \cdot \hat{e}_2 & \hat{e}_3 \cdot \hat{e}_3
\end{bmatrix} = \begin{bmatrix}
1 & 0 & 0 \\
0 & 1 & 0 \\
0 & 0 & 1
\end{bmatrix}, \tag{3.59}
$$

where we used

$$
\hat{x} \cdot \hat{x} = \hat{y} \cdot \hat{y} = \hat{z} \cdot \hat{z} = 1, \tag{3.60}
$$

and

$$
\hat{x} \cdot \hat{z} = \hat{z} \cdot \hat{x} = \hat{x} \cdot \hat{y} = \hat{y} \cdot \hat{x} = \hat{y} \cdot \hat{z} = \hat{z} \cdot \hat{y} = 0. \tag{3.61}
$$

Recalling that

$$
\Gamma^{\eta}_{\mu\beta} = \frac{g^{\eta\alpha}}{2}\left(\partial_\mu g_{\alpha\beta} + \partial_\beta g_{\mu\alpha} - \partial_\alpha g_{\beta\mu}\right),
$$
$$
\Rightarrow \Gamma^{\alpha}_{\mu\alpha} = \frac{g^{\alpha\alpha}}{2}\left(\partial_\mu g_{\alpha\alpha} + \partial_\alpha g_{\mu\alpha} - \partial_\alpha g_{\alpha\mu}\right), \tag{3.62}
$$

one finds from equation (3.59) $\Gamma^{\alpha}_{\mu\alpha} = 0$.

(c) In view of the result in part (b), the force in equation (3.54) becomes

$$
\vec{f} = \nabla \cdot \overset{\leftrightarrow}{T} = \partial_\alpha T^{\alpha\beta} + \Gamma^{\alpha}_{\mu\alpha} T^{\mu\beta} = \partial_\alpha T^{\alpha\beta}. \tag{3.63}
$$

Noting that this can be put in the form

$$
\vec{f} = [\partial_\alpha \hat{e}^\alpha] \cdot [T^{\alpha\beta}\hat{e}_\alpha \oplus \hat{e}_\alpha], \tag{3.64}
$$

we have

$$
\begin{aligned}
\vec{f} = [\partial_1 \hat{e}^1 &+ \partial_2 \hat{e}^2 + \partial_3 \hat{e}^3] \\
&\cdot \Bigg[\epsilon_0\left(E^1 E^1 - \frac{1}{2}EE\right)(\hat{e}_1 \oplus \hat{e}_1) + \epsilon_0 E^1 E^2 (\hat{e}_1 \oplus \hat{e}_2) + \epsilon_0 E^1 E^3 (\hat{e}_1 \oplus \hat{e}_3) \\
&\quad + \epsilon_0 E^2 E^1 (\hat{e}_2 \oplus \hat{e}_1) + \epsilon_0\left(E^2 E^2 - \frac{1}{2}EE\right)(\hat{e}_2 \oplus \hat{e}_2) + \epsilon_0 E^2 E^3 (\hat{e}_2 \oplus \hat{e}_3) \\
&\quad + \epsilon_0 E^3 E^1 (\hat{e}_3 \oplus \hat{e}_1) + \epsilon_0 E^3 E^2 (\hat{e}_3 \oplus \hat{e}_2) + \epsilon_0\left(E^3 E^3 - \frac{1}{2}EE\right)(\hat{e}_3 \oplus \hat{e}_3) \Bigg],
\end{aligned} \tag{3.65}
$$

which results in

$$
\begin{aligned}
\vec{f} = {}& \epsilon_0 \partial_1 \left(E^1 E^1 - \frac{1}{2} EE \right)[(\hat{e}_1 \cdot \hat{e}_1) \oplus \hat{e}_1] + \epsilon_0 \partial_1 (E^1 E^2)[(\hat{e}_1 \cdot \hat{e}_1) \oplus \hat{e}_2] \\
& + \epsilon_0 \partial_1 (E^1 E^3)[(\hat{e}_1 \cdot \hat{e}_1) \oplus \hat{e}_3] \\
& + \epsilon_0 \partial_2 (E^2 E^1)[(\hat{e}^2 \cdot \hat{e}_2) \oplus \hat{e}_1] + \epsilon_0 \partial_2 \left(E^2 E^2 - \frac{1}{2} EE \right)[(\hat{e}^2 \cdot \hat{e}_2) \oplus \hat{e}_2] \\
& + \epsilon_0 \partial_2 (E^2 E^3)[(\hat{e}^2 \cdot \hat{e}_2) \oplus \hat{e}_3] \\
& + \epsilon_0 \partial_3 (E^3 E^1)[(\hat{e}^3 \cdot \hat{e}_3) \oplus \hat{e}_1] + \epsilon_0 \partial_3 (E^3 E^2)[(\hat{e}^3 \cdot \hat{e}_3) \oplus \hat{e}_2] \\
& + \epsilon_0 \partial_3 \left(E^3 E^3 - \frac{1}{2} EE \right)[(\hat{e}^3 \cdot \hat{e}_3) \oplus \hat{e}_3].
\end{aligned}
\tag{3.66}
$$

Recalling that $\hat{e}^\alpha \cdot \hat{e}_\beta = \delta^\alpha_\beta$, we find

$$
\begin{aligned}
[f^\beta] = {}& \epsilon_0 \partial_1 \left(E^1 E^1 - \frac{1}{2} EE \right)\hat{e}_1 + \epsilon_0 \partial_1 (E^1 E^2)\hat{e}_2 + \epsilon_0 \partial_1 (E^1 E^3)\hat{e}_3 \\
& + \epsilon_0 \partial_2 (E^2 E^1)\hat{e}_1 + \epsilon_0 \partial_2 \left(E^2 E^2 - \frac{1}{2} EE \right)\hat{e}_2 + \epsilon_0 \partial_2 (E^2 E^3)\hat{e}_3 \\
& + \epsilon_0 \partial_3 (E^3 E^1)\hat{e}_1 + \epsilon_0 \partial_3 (E^3 E^2)\hat{e}_2 + \epsilon_0 \partial_3 \left(E^3 E^3 - \frac{1}{2} EE \right)\hat{e}_3,
\end{aligned}
\tag{3.67}
$$

which can be put in the form

$$
\begin{aligned}
\vec{f} = {}& \epsilon_0 \left[\partial_1 \left(E^1 E^1 - \frac{1}{2} EE \right) + \partial_2 (E^2 E^1) + \partial_3 (E^3 E^1) \right]\hat{e}_1 \\
& + \epsilon_0 \left[\partial_1 (E^1 E^2) + \partial_2 \left(E^2 E^2 - \frac{1}{2} EE \right) + \partial_3 (E^3 E^2) \right]\hat{e}_2 \\
& + \epsilon_0 \left[\partial_1 (E^1 E^3) + \partial_2 (E^2 E^3) + \partial_3 \left(E^3 E^3 - \frac{1}{2} EE \right) \right]\hat{e}_3.
\end{aligned}
\tag{3.68}
$$

Using

$$
EE = E^1 E^1 + E^2 E^2 + E^3 E^3,
\tag{3.69}
$$

equation (3.68) can be rewritten as

$$
\vec{f} = f^1 \hat{e}_1 + f^2 \hat{e}_2 + f^3 \hat{e}_3 = f^\alpha \hat{e}_\alpha,
\tag{3.70}
$$

where

$$
\begin{aligned}
f^1 &= \epsilon_0 \left[\frac{1}{2}\partial_1 (E^1 E^1 - E^2 E^2 - E^3 E^3) + \partial_2 (E^2 E^1) + \partial_3 (E^3 E^1) \right], \\
f^2 &= \epsilon_0 \left[\partial_1 (E^1 E^2) + \frac{1}{2}\partial_2 (-E^1 E^1 + E^2 E^2 - E^3 E^3) + \partial_3 (E^3 E^2) \right], \\
f^3 &= \epsilon_0 \left[\partial_1 (E^1 E^3) + \partial_2 (E^2 E^3) + \frac{1}{2}\partial_3 (-E^1 E^1 - E^2 E^2 + E^3 E^3) \right].
\end{aligned}
\tag{3.71}
$$

Note that the force $\vec{f}$ is a rank-1 tensor.

3.3 Permutations and symmetries in tensors

A tensor can be symmetric or antisymmetric. For a second-rank tensor, $t(\vec{u}, \vec{v})$:

$$t(\vec{u}, \vec{v}) = \begin{cases} - t(\vec{v}, \vec{u}), & \text{Antisymmetric,} \\ t(\vec{v}, \vec{u}), & \text{symmetric.} \end{cases} \tag{3.72}$$

Any tensor can be expressed as a sum of symmetric and antisymmetric tensors. Again, if we consider a second-rank tensor with elements $t_{\alpha\beta}$, we can express this elements as

$$t_{\alpha\beta} = \frac{1}{2}(t_{\alpha\beta} + t_{\beta\alpha}) + \frac{1}{2}(t_{\alpha\beta} - t_{\beta\alpha}). \tag{3.73}$$

Introducing the notations for the symmetric part,

$$t_{(\alpha\beta)} = \frac{1}{2}(t_{\alpha\beta} + t_{\beta\alpha}), \tag{3.74}$$

and the antisymmetric part,

$$t_{[\alpha\beta]} = \frac{1}{2}(t_{\alpha\beta} - t_{\beta\alpha}), \tag{3.75}$$

we can write

$$t_{\alpha\beta} = t_{(\alpha\beta)} + t_{[\alpha\beta]}. \tag{3.76}$$

Nth-rank tensor: For an Nth-rank tensor the symmetric and antisymmetric covariant components of the tensor, $t_{\alpha_1\alpha_2\cdots\alpha_N}$, are given by

$$t_{(\alpha_1\alpha_2\cdots\alpha_N)} = \frac{1}{N!}(\text{addition over all permutations of the indices } \alpha_1\alpha_2\cdots\alpha_N), \tag{3.77}$$

and

$$t_{[\alpha_1\alpha_2\cdots\alpha_N]} = \frac{1}{N!} \text{ (Alternatingsubtraction and addition over} \tag{3.78}$$
$$\text{all permutations of the indices } \alpha_1\alpha_2\cdots\alpha_N).$$

For example for third-rank tensor, we have

$$t_{(\alpha_1\alpha_2\alpha_3)} = \frac{1}{3!}(t_{\alpha_1\alpha_2\alpha_3} + t_{\alpha_2\alpha_1\alpha_3} + t_{\alpha_1\alpha_3\alpha_2} + t_{\alpha_3\alpha_1\alpha_2} + t_{\alpha_3\alpha_2\alpha_1} + t_{\alpha_2\alpha_3\alpha_1}), \tag{3.79}$$

and

$$t_{[\alpha_1\alpha_2\alpha_3]} = \frac{1}{3!}(t_{\alpha_1\alpha_2\alpha_3} - t_{\alpha_2\alpha_1\alpha_3} + t_{\alpha_1\alpha_3\alpha_2} - t_{\alpha_3\alpha_1\alpha_2} + t_{\alpha_3\alpha_2\alpha_1} - t_{\alpha_2\alpha_3\alpha_1}). \tag{3.80}$$

Particular subset of indices permutation: We use different notations when the permutation is applied to a particular subset of indices. This is described using a fourth-rank tensor:

symmetric permutation to indices α and β only

$$t_{(\alpha\beta)\kappa\mu} = \frac{1}{2}(t_{\alpha\beta\kappa\mu} + t_{\beta\alpha\kappa\mu}), \tag{3.81}$$

antisymmetric permutation to indices α and β only

$$t_{[\alpha\beta]\kappa\mu} = \frac{1}{2}(t_{\alpha\beta\kappa\mu} - t_{\beta\alpha\kappa\mu}), \tag{3.82}$$

antisymmetric permutation to indices β and μ only

$$t_{\alpha[\beta|\kappa|\mu]} = \frac{1}{2}(t_{\alpha\beta\kappa\mu} - t_{\alpha\mu\kappa\beta}), \tag{3.83}$$

symmetric permutation to indices β and μ only

$$t_{\alpha(\beta|\kappa|\mu)} = \frac{1}{2}(t_{\alpha\beta\kappa\mu} + t_{\alpha\mu\kappa\beta}). \tag{3.84}$$

Note that the $\|$ symbols are used to exclude unwanted indices from the symmetrization () or antisymmetrization [] implied.

3.4 Associated tensors

We have seen that a good example of a second-rank tensor is a metric tensor. We recall that covariant and contravariant components of the metric tensor are given by

$$g_{\alpha\beta} = g(\hat{e}_\alpha, \hat{e}_\beta) = \hat{e}_\alpha \cdot \hat{e}_\beta, \ g^{\alpha\beta} = g(\hat{e}^\alpha, \hat{e}^\beta) = \hat{e}^\alpha \cdot \hat{e}^\beta, \tag{3.85}$$

and the mixed components:

$$g_\alpha^{\ \beta} = g(\hat{e}_\alpha, \hat{e}^\beta) = g(\hat{e}^\alpha, \hat{e}_\beta) = \delta_\beta^{\ \alpha}. \tag{3.86}$$

Consider the third-rank tensor t, expressed in terms of its covariant components:

$$t_{\alpha\beta\kappa} = t(\hat{e}_\alpha, \hat{e}_\beta, \hat{e}_\kappa), \tag{3.87}$$

or mixed components:

$$t_{\alpha\beta}^{\ \ \kappa} = t(\hat{e}^\kappa, \hat{e}_\alpha, \hat{e}_\beta). \tag{3.88}$$

We recall that the covariant form of the metric tensor can be used to lower indices and the contravariant form can be used to raise indices. Thus to lower the index in equation (3.88), we multiply by the covariant form of the metric tensor $g_{\mu\kappa}$:

$$g_{\mu\kappa}t_{\alpha\beta}^{\ \ \kappa} = g(\hat{e}_\mu, \hat{e}_\kappa)t(\hat{e}^\kappa, \hat{e}_\alpha, \hat{e}_\beta) = t(\hat{e}_\mu, \hat{e}_\alpha, \hat{e}_\beta) = t_{\mu\alpha\beta}. \tag{3.89}$$

Similarly,

$$g^{\mu\alpha}t_{\alpha\beta\kappa} = g(\hat{e}^{\mu}, \hat{e}^{\alpha})t(\hat{e}_{\alpha}, \hat{e}_{\beta}, \hat{e}_{\kappa}) = t(\hat{e}^{\mu}, \hat{e}_{\beta}, \hat{e}_{\kappa}) = t^{\mu}_{\beta\kappa} \tag{3.90}$$

raises the index α of the tensor $t_{\alpha\beta\kappa}$.

3.5 Mapping tensors onto tensors

We have seen that tensors map vectors into real numbers. A rank-1 tensor, $t(\vec{u})$, maps the vector $\vec{u}$ into real numbers in the tangent space

$$t(\vec{u}) = t(u^{\alpha}\hat{e}_{\alpha}) = u^{\alpha}t(\hat{e}_{\alpha}) = t_{\alpha}u^{\alpha}, \tag{3.91}$$

and the second-rank tensor, $t(\vec{u}, \vec{v})$, maps the vectors $\vec{u}$ and $\vec{v}$ into real numbers in the tangent space

$$t(\vec{u}, \vec{v}) = t(u^{\alpha}\hat{e}_{\alpha}, v^{\beta}\hat{e}_{\beta}) = t(\hat{e}_{\alpha}, \hat{e}_{\beta})u^{\alpha}v^{\beta} = t_{\alpha\beta}u^{\alpha}v^{\beta}. \tag{3.92}$$

Now, the question is whether we can map a tensor into another tensor of a different rank. Consider a rank-3 tensor $t(\vec{u}, \vec{v}, \vec{w})$, which maps the three vectors into real numbers. If we replace the two vectors $(\vec{u}, \vec{v})$ by the basis vectors $(\hat{e}_{\alpha}, \hat{e}_{\beta})$, we have

$$t(\hat{e}_{\alpha}, \hat{e}_{\beta}, \vec{w}) = (\hat{e}_{\alpha} \cdot \hat{e}_{\beta}) \cdot \vec{w}, \tag{3.93}$$

so that using the contravariant components for $\vec{w}$,

$$\vec{w} = w^{\kappa}\hat{e}, \tag{3.94}$$

we find

$$t(\hat{e}_{\alpha}, \hat{e}_{\beta}, \vec{w}) = (\hat{e}_{\alpha} \cdot \hat{e}_{\beta}) \cdot (w^{\kappa}\hat{e}_{\kappa}) = (\hat{e}_{\alpha} \cdot \hat{e}_{\beta} \cdot \hat{e}_{\kappa})w^{\kappa}$$
$$\Rightarrow t(\hat{e}_{\alpha}, \hat{e}_{\beta}, \vec{w}) = t_{\alpha\beta\kappa}w^{\kappa} = s_{\alpha\beta}, \tag{3.95}$$

or using the covariant component,

$$\vec{w} = w_{\kappa}\hat{e}^{\kappa},$$

we have

$$t(\hat{e}_{\alpha}, \hat{e}_{\beta}, \vec{w}) = (\hat{e}_{\alpha} \cdot \hat{e}_{\beta}) \cdot (w_{\kappa}\hat{e}^{\kappa}) = (\hat{e}_{\alpha} \cdot \hat{e}_{\beta} \cdot \hat{e}^{\kappa})w_{\kappa}$$
$$\Rightarrow t(\hat{e}_{\alpha}, \hat{e}_{\beta}, \vec{w}) = t^{\kappa}_{\alpha\beta}w_{\kappa} = s_{\alpha\beta}, \tag{3.96}$$

which is a rank-2 tensor. In terms of the metric tensor, one can also find this second-rank tensor,

$$t(\hat{e}_{\alpha}, \hat{e}_{\beta}, \vec{w}) = (\hat{e}_{\alpha} \cdot \hat{e}_{\beta}) \cdot (w^{\kappa}\hat{e}_{\kappa}) = (\hat{e}_{\alpha} \cdot \hat{e}_{\kappa}) \cdot (w^{\kappa}\hat{e}_{\beta})$$
$$\Rightarrow t(\hat{e}_{\alpha}, \hat{e}_{\beta}, \vec{w}) = g_{\alpha\kappa}w^{\kappa} \cdot \hat{e}_{\kappa} = [w_{\alpha}] \cdot \hat{e}_{\kappa} = s_{\alpha\kappa} \tag{3.97}$$

or

$$t(\hat{e}_\alpha, \hat{e}_\beta, \vec{w}) = (\hat{e}_\alpha \cdot \hat{e}_\beta) \cdot (w_\kappa \hat{e}^\kappa) = \hat{e}_\alpha \cdot \left[(\hat{e}_\beta \cdot \hat{e}^\kappa) w_\kappa \right]$$
$$\Rightarrow t(\hat{e}_\alpha, \hat{e}_\beta, \vec{w}) = \hat{e}_\alpha \cdot \left[\delta_\beta^\kappa w_\kappa \right] = \hat{e}_\alpha \cdot [w_\beta] = s_{\alpha\beta}. \tag{3.98}$$

The results in equations (3.97) and (3.98) show that a rank-3 tensor, t, maps the vector, $\vec{w}$, into a rank-2 tensor, s. As another example, let us consider a rank-3 tensor defined by

$$t(\vec{u}, \hat{e}_\beta, \vec{w}) = \vec{v} \cdot \hat{e}_\beta \cdot \vec{w}. \tag{3.99}$$

Using the vectors $\vec{v}$ and $\vec{u}$,

$$\vec{w} = w^\kappa \hat{e}_\kappa = w_\kappa \hat{e}^\kappa, \ \vec{u} = u^\alpha \hat{e}_\alpha = u_\alpha \hat{e}^\alpha, \tag{3.100}$$

one can write

$$t(\vec{u}, \hat{e}_\beta, \vec{w}) = u^\alpha \hat{e}_\alpha \cdot \hat{e}_\beta \cdot w^\kappa \hat{e}_\kappa = (\hat{e}_\alpha \cdot \hat{e}_\beta \cdot \hat{e}_\kappa) u^\alpha w^\kappa$$
$$\Rightarrow t(\vec{u}, \hat{e}_\beta, \vec{w}) = t_{\alpha\beta\kappa} u^\alpha w^\kappa = s_\beta, \tag{3.101}$$

$$t(\vec{u}, \hat{e}_\beta, \vec{w}) = u_\alpha \hat{e}^\alpha \cdot \hat{e}_\beta \cdot w^\kappa \hat{e}_\kappa = (\hat{e}^\alpha \cdot \hat{e}_\beta \cdot \hat{e}_\kappa) u_\alpha w^\kappa$$
$$\Rightarrow t(\vec{u}, \hat{e}_\beta, \vec{w}) = t_{\beta\kappa}^\alpha u_\alpha w^\kappa = s_\beta, \tag{3.102}$$

$$t(\vec{u}, \hat{e}_\beta, \vec{w}) = u^\alpha \hat{e}_\alpha \cdot \hat{e}_\beta \cdot w_\kappa \hat{e}^\kappa = (\hat{e}_\alpha \cdot \hat{e}_\beta \cdot \hat{e}^\kappa) u^\alpha w_\kappa$$
$$\Rightarrow t(\vec{u}, \hat{e}_\beta, \vec{w}) = t_{\alpha\beta}^\kappa u^\alpha w_\kappa = s_\beta, \tag{3.103}$$

$$t(\vec{u}, \hat{e}_\beta, \vec{w}) = u_\alpha \hat{e}^\alpha \cdot \hat{e}_\beta \cdot w_\kappa \hat{e}^\kappa = (\hat{e}^\alpha \cdot \hat{e}_\beta \cdot \hat{e}^\kappa) u_\alpha w_\kappa$$
$$\Rightarrow t(\vec{u}, \hat{e}_\beta, \vec{w}) = t_\beta^{\alpha\kappa} u_\alpha w_\kappa = s_\beta, \tag{3.104}$$

where we find a rank-1 tensor, s_β. This shows that a rank-3 tensor mapped the two vectors into a rank-1 tensor.

Example 3.4. Using the two-dimensional vectors

$$\vec{u} = u^1 \hat{e}_1 + u^2 \hat{e}_2, \ \vec{w} = w^1 \hat{e}_1 + w^2 \hat{e}_2. \tag{3.105}$$

(a) Show $t(\hat{e}_\alpha, \hat{e}_\beta, \vec{w})$ maps the vector $\vec{w}$ into a second-rank tensor.
(b) Show that $t(\vec{u}, \hat{e}_\beta, \vec{w})$ maps the two-vector into a rank-1 tensor.
Solution:
  (a) Using the result in equation (3.97) one can also find this second-rank tensor,

$$t(\hat{e}_\alpha, \hat{e}_\beta, \vec{w}) = [w_\alpha] \cdot \hat{e}_\kappa = s_{\alpha\kappa},$$

and one can write

$$[w_1] \cdot \hat{e}_1 + [w_1] \cdot \hat{e}_2 + [w_2] \cdot \hat{e}_1 + [w_2] \cdot \hat{e}_2,$$

which is a rank-two tensor.

(b) Using the metric tensor, one can write

$$t(\vec{u}, \hat{e}_\beta, \vec{w}) = u^\alpha \hat{e}_\alpha \cdot \hat{e}_\beta \cdot w^\kappa \hat{e}_\kappa = (\hat{e}_\alpha \cdot \hat{e}_\beta) u^\alpha \cdot w^\kappa \hat{e}_\kappa$$
$$= \left[g_{\alpha\beta} u^\alpha \right] \cdot w^\kappa \hat{e}_\kappa \Rightarrow t(\vec{u}, \hat{e}_\beta, \vec{w}) = [u_\beta] \cdot w^\kappa \hat{e}_\kappa. \tag{3.106}$$

For the vectors

$$\vec{u} = u^1 \hat{e}_1 + u^2 \hat{e}_2, \; \vec{w} = w^1 \hat{e}_1 + w^2 \hat{e}_2, \tag{3.107}$$

and one can write

$$t(\vec{u}, \hat{e}_\beta, \vec{w}) = [u_\beta] \cdot w^\kappa \hat{e}_\kappa$$
$$= [u_1] \cdot (w^1 \hat{e}_1 + w^2 \hat{e}_2) + [u_2] \cdot (w^1 \hat{e}_1 + w^2 \hat{e}_2) \tag{3.108}$$
$$= ([u_1 + u_2] \cdot w^1 \hat{e}_1 + [u_1 + u_2] \cdot w^2 \hat{e}_2),$$

which is a rank-1 tensor.

Tensors and inner product

From what we know up to this point, the inner product of two rank-1 tensors (two vectors) is commutative. However, that generalization does not apply to higher-ranked tensors. So from now on, we must keep in mind that the inner product of tensors, in general, is not commutative, including the first-rank tensor, for the reason that we detail below in tensor contraction. Let us consider a rank-1 tensor, $\vec{s}$, and a rank-2 tensor, $\overleftrightarrow{t}$. The inner product of these two tensors can be written as

$$\overleftrightarrow{t} \cdot \vec{s} = t^{\alpha\beta} s_\beta, \tag{3.109}$$

and

$$\vec{s} \cdot \overleftrightarrow{t} = s_\alpha t^{\alpha\beta}. \tag{3.110}$$

These two equations are not necessarily the same. For example, when these tensors are in a two-dimensional tangent space, we can write

$$\overleftrightarrow{t} \cdot \vec{s} = t^{\alpha\beta} s_\beta = \begin{bmatrix} t^{11} & t^{12} \\ t^{21} & t^{22} \end{bmatrix} \begin{bmatrix} s_1 \\ s_2 \end{bmatrix}$$
$$\Rightarrow \vec{s} \cdot \overleftrightarrow{t} = t^{\alpha\beta} s_\beta = \begin{bmatrix} s_1 t^{11} + s_2 t^{12} \\ s_1 t^{21} + s_2 t^{22} \end{bmatrix}, \tag{3.111}$$

and

$$\vec{s} \cdot \overleftrightarrow{t} = s_\alpha t^{\alpha\beta} = \begin{bmatrix} s_1 & s_2 \end{bmatrix} \begin{bmatrix} t^{11} & t^{12} \\ t^{21} & t^{22} \end{bmatrix}$$
$$\Rightarrow \vec{s} \cdot \overleftrightarrow{t} = s_\alpha t^{\alpha\beta} = \begin{bmatrix} s_1 t^{11} + s_2 t^{21} & s_1 t^{12} + s_2 t^{22} \end{bmatrix}. \tag{3.112}$$

The two results in equations (3.111) and (3.112) are equal only when the rank-2 tensor is symmetric, $t^{12} = t^{21}$. Therefore, the inner product is not commutative if it involves rank-2 and higher tensors.

Tensors are independent of representations
We already know that first-rank tensors (vectors) are geometrical objects that we can construct from a linear combination of basis vectors:

$$\vec{t} = t_\alpha \hat{e}^\alpha = t^\alpha \hat{e}_\alpha. \tag{3.113}$$

The vector that defines a given geometry on a manifold does not depend on how we represent it. The geometry that a vector defines remains the same whatever representation we use to describe the vector. The same is true for higher-ranked tensors. Consider a rank-2 tensor constructed from the outer product of two basis vectors of some coordinate system:

$$\overset{\leftrightarrow}{w} = \hat{e}_\alpha \otimes \hat{e}_\beta. \tag{3.114}$$

Then the contravariant component of this tensor is given by

$$w^{\alpha\beta} = \hat{e}_\alpha \otimes \hat{e}_\beta(\hat{e}^\kappa, \hat{e}^\mu) = \hat{e}_\alpha(\hat{e}^\kappa) \otimes \hat{e}_\beta(\hat{e}^\mu) = \delta_\alpha^\kappa \delta_\beta^\mu. \tag{3.115}$$

Now, suppose we have some general second-rank tensor, $\overset{\leftrightarrow}{t}$, with a contravariant component, $t^{\alpha\beta}$. Then the quantity

$$t^{\alpha\beta} w_{\alpha\beta} = t^{\alpha\beta}\left(\hat{e}_\alpha \otimes \hat{e}_\beta\right) \tag{3.116}$$

represents the sum of two second-rank tensors, which results in a rank-2 tensor. The action of this second-rank tensor on two basis vectors results in

$$t^{\alpha\beta}\left(\hat{e}_\alpha \otimes \hat{e}_\beta\right)(\hat{e}^\kappa, \hat{e}^\mu) = t^{\alpha\beta}\delta_\alpha^\kappa \delta_\beta^\mu = t^{\kappa\mu}, \tag{3.117}$$

which is the contravariant component of the rank-2 tensor $\overset{\leftrightarrow}{t}$. The result in equation (3.117) shows that

$$\overset{\leftrightarrow}{t} = t^{\alpha\beta}\left(\hat{e}_\alpha \otimes \hat{e}_\beta\right) = t_\alpha^\beta\left(\hat{e}^\alpha \otimes \hat{e}_\beta\right) = t_\beta^\alpha(\hat{e}_\alpha \otimes \hat{e}^\beta).$$

This is true for any higher-ranked tensor.

3.6 Tensors and coordinate transformations

We recall the coordinate basis and dual basis vectors, under coordinate transformation, $x^\alpha \to x'^\alpha$, are determined from

$$\hat{e}'_\alpha = \frac{\partial x^\kappa}{\partial x'^\alpha}\hat{e}_\kappa, \ \hat{e}'^\alpha = \frac{\partial x'^\alpha}{\partial x^\kappa}\hat{e}^\kappa. \tag{3.118}$$

Rank-1 tensor
We have already seen how a first-rank tensor transforms under coordinate transformation in the previous chapter. We saw that, in the x^α coordinates,

$$\vec{t} = t_\alpha \hat{e}^\alpha = t^\alpha \hat{e}_\alpha, \tag{3.119}$$

and in the $x^{'\alpha}$ coordinates,

$$\vec{t}' = t_\alpha' \hat{e}^{'\alpha} = t^{'\alpha} \hat{e}_\alpha', \tag{3.120}$$

are equal:

$$\vec{t} = \vec{t}'. \tag{3.121}$$

Using this property, we have shown that

$$t^{'\alpha} = \vec{t}'\left(\hat{e}_\beta'\right) \cdot \hat{e}^{'\alpha} = \vec{t}\left(\hat{e}_\beta\right) \cdot \hat{e}^{'\alpha} = t^\beta \hat{e}_\beta \cdot \hat{e}^{'\alpha}$$

$$= t^\beta \hat{e}_\beta \cdot \frac{\partial x^{'\alpha}}{\partial x^\kappa} \hat{e}^\kappa = \frac{\partial x^{'\alpha}}{\partial x^\kappa} t^\beta \hat{e}_\beta \cdot \hat{e}^\kappa = \frac{\partial x^{'\alpha}}{\partial x^\kappa} t^\beta \delta_\beta^\kappa \tag{3.122}$$

$$\Rightarrow t^{'\alpha} = \frac{\partial x^{'\alpha}}{\partial x^\kappa} t^\kappa,$$

and

$$t_\alpha' = \vec{t}'(\hat{e}^{'\beta}) \cdot \hat{e}_\alpha' = \vec{t}(\hat{e}^\beta) \cdot \hat{e}_\alpha' = t_\beta \hat{e}^\beta \cdot \hat{e}^{'\alpha}$$

$$= t_\beta \hat{e}^\beta \cdot \frac{\partial x^\kappa}{\partial x^{'\alpha}} \hat{e}_\kappa = \frac{\partial x^\kappa}{\partial x^{'\alpha}} t_\beta \hat{e}^\beta \cdot \hat{e}_\kappa = \frac{\partial x^\kappa}{\partial x^{'\alpha}} t_\beta \delta_\kappa^\beta \tag{3.123}$$

$$\Rightarrow t_\alpha' = \frac{\partial x^\kappa}{\partial x^{'\alpha}} t_\kappa.$$

Rank-2 tensor

Consider a rank-2 tensor $\overset{\leftrightarrow}{t}$ in the x^α coordinates:

$$\overset{\leftrightarrow}{t} = t_{\alpha\beta}(\hat{e}^\alpha \otimes \hat{e}^\beta) = t^{\alpha\beta}\left(\hat{e}_\alpha \otimes \hat{e}_\beta\right) = t_\beta^\alpha(\hat{e}_\alpha \otimes \hat{e}^\alpha), \tag{3.124}$$

which becomes in the $x^{'\alpha}$ coordinates

$$\overset{\leftrightarrow}{t}' = t_{\alpha\beta}'(\hat{e}^{'\alpha} \otimes \hat{e}^{'\beta}) = t^{'\alpha\beta}\left(\hat{e}_\alpha' \otimes \hat{e}_\beta'\right) = t_\beta^{'\alpha}(\hat{e}_\alpha' \otimes \hat{e}^\beta'), \tag{3.125}$$

where

$$\overset{\leftrightarrow}{t} = \overset{\leftrightarrow}{t}'. \tag{3.126}$$

Noting that

$$t_{\alpha\beta}' = \overset{\leftrightarrow}{t}'(\hat{e}^{'\alpha} \otimes \hat{e}^{'\beta}) \cdot \left(\hat{e}_\kappa', \hat{e}_\mu'\right) = \overset{\leftrightarrow}{t}(\hat{e}^\alpha \otimes \hat{e}^\beta) \cdot \left(\hat{e}_\kappa', \hat{e}_\mu'\right)$$

$$= t_{\alpha\beta}(\hat{e}^\alpha \otimes \hat{e}^\beta) \cdot \left(\hat{e}_\kappa', \hat{e}_\mu'\right) = t_\alpha(\hat{e}^\alpha) \cdot (\hat{e}_\kappa') \otimes t_\beta(\hat{e}^\beta) \cdot \left(\hat{e}_\mu'\right), \tag{3.127}$$

where we expressed

$$t_{\alpha\beta}(\hat{e}^\alpha \otimes \hat{e}^\beta) = t_\alpha(\hat{e}^\alpha) \otimes t_\beta(\hat{e}^\beta). \tag{3.128}$$

Applying the transformation for the basis vectors,

$$\hat{e}'_\kappa = \frac{\partial x^\nu}{\partial x'^\kappa}\hat{e}_\nu, \ \hat{e}'_\mu = \frac{\partial x^\eta}{\partial x'^\mu}\hat{e}_\eta, \tag{3.129}$$

we find

$$t'_{\alpha\beta} = t_\alpha(\hat{e}^\alpha) \cdot \left(\frac{\partial x^\nu}{\partial x'^\kappa}\hat{e}_\nu\right) \otimes t_\beta(\hat{e}^\beta) \cdot \left(\frac{\partial x^\eta}{\partial x'^\mu}\hat{e}_\eta\right),$$

$$= \frac{\partial x^\nu}{\partial x'^\kappa}\frac{\partial x^\eta}{\partial x'^\mu}(t_\alpha(\hat{e}^\alpha) \cdot \hat{e}_\nu) \otimes (t_\beta(\hat{e}^\beta) \cdot \hat{e}_\eta) = \frac{\partial x^\nu}{\partial x'^\kappa}\frac{\partial x^\eta}{\partial x'^\mu}t_{\alpha\beta}\delta^\alpha_\nu\delta^\beta_\eta \tag{3.130}$$

$$\Rightarrow t'_{\alpha\beta} = t'\left(\hat{e}'_\alpha, \hat{e}'_\beta\right) = \frac{\partial x^\alpha}{\partial x'^\kappa}\frac{\partial x^\beta}{\partial x'^\mu}t_{\alpha\beta}.$$

Following a similar procedure one can show that for the contravariant and mixed components,

$$t'^{\alpha\beta} = t'\left(\hat{e}'_\alpha, \hat{e}'_\beta\right) = \frac{\partial x'^\alpha}{\partial x^\kappa}\frac{\partial x'^\beta}{\partial x^\mu}t^{\kappa\mu}, \tag{3.131}$$

$$t'^\beta_\alpha = t'(\hat{e}'_\alpha, \hat{e}^{\beta'}) = \frac{\partial x^\kappa}{\partial x'^\alpha}\frac{\partial x'^\beta}{\partial x^\mu}t^\mu_\kappa. \tag{3.132}$$

Rank-3 tensor

Suppose we have a mixed third-rank tensor, $t^\kappa_{\alpha\beta} \to t\left(\hat{e}_\alpha, \hat{e}_\beta, \hat{e}^\kappa\right)$. In the x'^α coordinate system this is given by

$$t'^\kappa_{\alpha\beta} = t\left(\hat{e}'_\alpha, \hat{e}'_\beta, \hat{e}'^\kappa\right) = \frac{\partial x^\mu}{\partial x'^\alpha}\frac{\partial x^\nu}{\partial x'^\beta}\frac{\partial x'^\kappa}{\partial x^f}t^f_{\mu\nu}. \tag{3.133}$$

See problem 7.

3.7 Tensor equations and the quotient theorem

A tensor equation that holds in one coordinate system must hold in another coordinate system. Suppose we have an equation that states two rank-2 tensors, $\overset{\leftrightarrow}{t}$ and $\overset{\leftrightarrow}{s}$, are equal in the x^α coordinate system. That means

$$t_{\alpha\beta} = s_{\alpha\beta}. \tag{3.134}$$

Multiplying both sides of this equation by

$$\frac{\partial x^\alpha}{\partial x'^\kappa}\frac{\partial x^\beta}{\partial x'^\mu}, \tag{3.135}$$

we have

$$\frac{\partial x^\alpha}{\partial x'^\kappa}\frac{\partial x^\beta}{\partial x'^\mu}t_{\alpha\beta} = \frac{\partial x^\alpha}{\partial x'^\kappa}\frac{\partial x^\beta}{\partial x'^\mu}s_{\alpha\beta}, \tag{3.136}$$

so that applying equation (3.130), we find

$$t'_{\alpha\beta} = s'_{\alpha\beta}. \tag{3.137}$$

Equation (3.130) shows that the equality holds under the coordinate transformation. However, the question is whether these components (i.e., a set of quantities) are actually form the components of a tensor. The *quotient theorem* sets the condition for a set of quantities to be the components of a tensor.

The quotient theorem

The quotient theorem states that for a set of quantities (components) to be a tensor, when the quantities of the tensor are contracted the tensor must produce another tensor under any coordinate transformation.

Suppose in an N-dimensional manifold you are given a rank-3 tensor, t, and a rank-1 tensor, v. The tensor t has a set of N^3 quantities (components), $t^\alpha_{\beta\kappa}$, and the tensor v has N quantities of v^α. We form a rank-4 tensor, s, by taking the outer product of these two tensors:

$$s^{\alpha\mu}_{\beta\kappa} = t^\alpha_{\beta\kappa} v^\mu.$$

We then form a set of N^2 components (quantities) by contracting the rank-4 tensor, s,

$$s^\alpha_\beta = s^{\alpha\kappa}_{\beta\kappa} = t^\alpha_{\beta\kappa} v^\kappa. \tag{3.138}$$

Under coordinate transformation $x^\alpha \to x'^\alpha$, this equation becomes

$$s'^\alpha_\beta = t'^\alpha_{\beta\kappa} v'^\kappa. \tag{3.139}$$

Applying the relation in equations (3.122) and (3.123), one can write

$$s'^\alpha = \frac{\partial x'^\alpha}{\partial x^\mu} s^\mu, \quad s'_\beta = \frac{\partial x^\nu}{\partial x'^\beta} s_\nu, \tag{3.140}$$

so that

$$s'^\alpha_\beta = \frac{\partial x^\nu}{\partial x'^\beta} s_\nu \frac{\partial x'^\alpha}{\partial x^\mu} s^\mu = \frac{\partial x^\nu}{\partial x'^\beta} \frac{\partial x'^\alpha}{\partial x^\mu} s^\mu_\nu. \tag{3.141}$$

Since s'^α_β is a rank-2 tensor and it must be transformed as a rank-2 tensor, one must have

$$s'^\alpha_\beta = \frac{\partial x^\nu}{\partial x'^\beta} \frac{\partial x'^\alpha}{\partial x^\mu} t^\mu_{\nu\eta} v^\eta. \tag{3.142}$$

Substituting the relation

$$v^\eta = \frac{\partial x^\eta}{\partial x'^\kappa} v'^\kappa \tag{3.143}$$

into equation (3.142), one finds

$$s_\beta^{'\alpha} = \frac{\partial x^\nu}{\partial x^{'\beta}} \frac{\partial x^{'\alpha}}{\partial x^\mu} \frac{\partial x^\eta}{\partial x^{'\kappa}} t_{\nu\eta}^{\mu} v^{'\kappa}.$$ (3.144)

Now, using equation (3.141) one can then rewrite equation (3.138) as

$$s_\beta^{'\alpha} = t_{\beta\kappa}^{'\alpha} v^{'\kappa} \Rightarrow \frac{\partial x^{'\alpha}}{\partial x^\mu} \frac{\partial x^\nu}{\partial x^{'\beta}} \frac{\partial x^\eta}{\partial x^{'\kappa}} t_{\nu\eta}^{\mu} v^{'\kappa} = t_{\beta\kappa}^{'\alpha} v^{'\kappa}$$

$$\Rightarrow \left(t_{\beta\kappa}^{'\alpha} - \frac{\partial x^{'\alpha}}{\partial x^\mu} \frac{\partial x^\nu}{\partial x^{'\beta}} \frac{\partial x^\eta}{\partial x^{'\kappa}} t_{\nu\eta}^{\mu} \right) v^{'\kappa} = 0.$$ (3.145)

There follows that for an arbitrary vector component $v^{'\kappa}$,

$$t_{\beta\kappa}^{'\alpha} = \frac{\partial x^{'\alpha}}{\partial x^\mu} \frac{\partial x^\nu}{\partial x^{'\beta}} \frac{\partial x^\eta}{\partial x^{'\kappa}} t_{\nu\eta}^{\mu}.$$ (3.146)

The result in equation (3.146) is in agreement with equation (3.133). This proves the quotient theorem. We were given a set of quantities (components) defined by $s_{\beta\kappa}^{\alpha\mu} = t_{\beta\kappa}^{\alpha} v^\mu$, of which we did not know whether the quantities were components of a rank-4 tensor. In order to determine this, we made a contraction, defined by $t_\beta^\alpha = s_{\beta\kappa}^{\alpha\kappa} = t_{\beta\kappa}^{\alpha} v^\kappa$, and still found a tensor verified by equation (3.146). Therefore, the set of elements $s_{\beta\kappa}^{\alpha\mu} = t_{\beta\kappa}^{\alpha} v^\mu$ must belong to the component of a rank-4 tensor.

3.8 Covariant derivatives of a tensor

Consider the component of a rank-1 tensor $v^{'\alpha}$ in the $x^{'\alpha}$ coordinates. The derivative of this tensor can be expressed as

$$\frac{\partial v^{'\alpha}}{\partial x^{'\beta}} = \frac{\partial v^{'\alpha}}{\partial x^\kappa} \frac{\partial x^\kappa}{\partial x^{'\beta}} = \frac{\partial x^\kappa}{\partial x^{'\beta}} \frac{\partial v^{'\alpha}}{\partial x^\kappa}.$$ (3.147)

Using the transformation for a rank-1 tensor,

$$v^{'\alpha} = \frac{\partial x^{'\alpha}}{\partial x^\mu} v^\mu,$$ (3.148)

one can write

$$\frac{\partial v^{'\alpha}}{\partial x^{'\beta}} = \frac{\partial x^\kappa}{\partial x^{'\beta}} \frac{\partial}{\partial x^\kappa} \left(\frac{\partial x^{'\alpha}}{\partial x^\mu} v^\mu \right) = \frac{\partial x^\kappa}{\partial x^{'\beta}} \frac{\partial x^{'\alpha}}{\partial x^\mu} \frac{\partial v^\kappa}{\partial x^\kappa} + \frac{\partial x^\kappa}{\partial x^{'\beta}} \frac{\partial^2 x^{'\alpha}}{\partial x^\kappa \partial x^\mu} v^\kappa.$$ (3.149)

In chapter 2 we saw that the derivative of a vector,

$$\vec{v} = v^\beta \hat{e}_\beta,$$

with respect to the coordinate x^α is given by

$$\partial_\alpha \vec{v} = \partial_\alpha(v^\beta \hat{e}_\beta) = (\partial_\alpha v^\beta + \Gamma_{\mu\alpha}^\beta v^\mu)\hat{e}_\beta = \nabla_\alpha v^\beta \hat{e}_\beta,$$ (3.150)

where the quantity

$$\nabla_\beta v^\alpha = \partial_\beta v^\alpha + \Gamma^\alpha_{\mu\beta} v^\mu \tag{3.151}$$

is the *covariant derivative* of the vector components, which is a mixed rank-2 tensor. Noting

$$\nabla = \hat{e}^\alpha \partial_\alpha, \ \vec{v} = v^\beta \hat{e}_\beta, \tag{3.152}$$

we can express the gradient of a vector as a tensor product of the gradient (rank-1 tensor) and the vector $\vec{v}$, which is also a rank-1 tensor:

$$\nabla\vec{v} = \hat{e}^\alpha \partial_\alpha \otimes \left(v^\beta \hat{e}_\beta \right) = \hat{e}^\alpha \otimes \partial_\alpha \left(v^\beta \hat{e}_\beta \right). \tag{3.153}$$

Applying equation (3.150), one can put equation (3.153) in the form

$$\nabla\vec{v} = (\nabla_\alpha v^\beta)\hat{e}^\alpha \otimes \hat{e}_\beta. \tag{3.154}$$

Let us consider the covariant derivative of a rank-2 tensor, $\overleftrightarrow{t}$, expressed in terms of its contravariant components $t^{\alpha\beta}$:

$$\nabla_\kappa \overleftrightarrow{t} = \nabla_\kappa t^{\alpha\beta} \hat{e}_\alpha \otimes \hat{e}_\beta. \tag{3.155}$$

Using the product rule, we have

$$\partial_\kappa \overleftrightarrow{t} = (\partial_\kappa t^{\alpha\beta})\hat{e}_\alpha \otimes \hat{e}_\beta + t^{\alpha\beta}(\partial_\kappa \hat{e}_\alpha) \otimes \hat{e}_\beta + t^{\alpha\beta}\hat{e}_\alpha \otimes \left(\partial_\kappa \hat{e}_\beta \right), \tag{3.156}$$

and applying the relation

$$\partial_\kappa \hat{e}_\beta = \Gamma^\eta_{\beta\kappa} \hat{e}_\eta, \tag{3.157}$$

one can write

$$\partial_\kappa \overleftrightarrow{t} = \left(\partial_\kappa t^{\alpha\beta} \right)\hat{e}_\alpha \otimes \hat{e}_\beta + t^{\alpha\beta}\Gamma^\eta_{\kappa\alpha}\hat{e}_\eta \otimes \hat{e}_\beta + t^{\alpha\beta}\hat{e}_\alpha \otimes \Gamma^\eta_{\kappa\beta}\hat{e}_\eta$$
$$\partial_\kappa \overleftrightarrow{t} = \left(\partial_\kappa t^{\alpha\beta} \right)\left(\hat{e}_\alpha \otimes \hat{e}_\beta \right) + t^{\alpha\beta}\Gamma^\eta_{\kappa\alpha}\left(\hat{e}_\eta \otimes \hat{e}_\beta \right) \tag{3.158}$$
$$+ t^{\alpha\beta}\Gamma^\eta_{\kappa\beta}\left(\hat{e}_\alpha \otimes \hat{e}_\eta \right).$$

If we interchange the indices η and α in the second term and η and β in the third term, we have

$$\partial_\kappa \overleftrightarrow{t} = (\partial_\kappa t^{\alpha\beta})\left(\hat{e}_\alpha \otimes \hat{e}_\beta \right) + t^{\eta\beta}\Gamma^\alpha_{\kappa\eta}\left(\hat{e}_\alpha \otimes \hat{e}_\beta \right)$$
$$+ t^{\alpha\eta}\Gamma^\beta_{\kappa\eta}\left(\hat{e}_\alpha \otimes \hat{e}_\beta \right), \tag{3.159}$$

which can be rewritten as

$$\partial_\kappa \overleftrightarrow{t} = \left[\left(\partial_\kappa t^{\alpha\beta} + \Gamma^\alpha_{\kappa\eta} t^{\eta\beta} + \Gamma^\beta_{\kappa\eta} t^{\alpha\eta} \right) \right]\left(\hat{e}_\alpha \otimes \hat{e}_\beta \right) \tag{3.160}$$

$$\Rightarrow \partial_\kappa \overleftrightarrow{t} = (\nabla_\kappa t^{\alpha\beta})\left(\hat{e}_\alpha \otimes \hat{e}_\beta \right), \tag{3.161}$$

where

$$\nabla_\kappa t^{\alpha\beta} = \partial_\kappa t^{\alpha\beta} + \Gamma^\alpha_{\kappa\mu} t^{\mu\beta} + \Gamma^\beta_{\kappa\mu} t^{\alpha\mu} \tag{3.162}$$

is the covariant derivative for a rank-2 tensor, which is clearly a rank-3 tensor.

Example 3.5. Show that the covariant derivative of the metric tensor,

$$\nabla_\kappa g^{\alpha\beta} = 0. \tag{3.163}$$

Solution: Using equation (3.162) one can write

$$\nabla_\kappa g^{\alpha\beta} = \partial_\kappa g^{\alpha\beta} + \Gamma^\alpha_{\kappa\mu} g^{\mu\beta} + \Gamma^\beta_{\kappa\mu} g^{\alpha\mu}. \tag{3.164}$$

Using the relation from chapter 2,

$$\partial_\kappa g^{\alpha\beta} = \partial_\kappa(\hat{e}^\alpha \cdot \hat{e}^\beta) = \partial_\kappa(\hat{e}^\alpha) \cdot \hat{e}^\beta + \hat{e}^\alpha \cdot \partial_\kappa(\hat{e}^\beta), \tag{3.165}$$

and

$$\partial_\kappa \hat{e}^\alpha = -\Gamma^\alpha_{\mu\kappa} \hat{e}^\mu, \ \partial_\kappa(\hat{e}^\beta) = -\Gamma^\beta_{\mu\kappa} \hat{e}^\mu, \tag{3.166}$$

one can write

$$\partial_\kappa g^{\alpha\beta} = -\Gamma^\alpha_{\mu\kappa} \hat{e}^\mu \cdot \hat{e}^\beta - \Gamma^\beta_{\mu\kappa} \hat{e}^\mu \cdot \hat{e}^\alpha = -\Gamma^\alpha_{\mu\kappa} g^{\mu\beta} - \Gamma^\beta_{\mu\kappa} g^{\mu\alpha}, \tag{3.167}$$

which leads to

$$\nabla_\kappa g^{\alpha\beta} = -\Gamma^\alpha_{\mu\kappa} g^{\mu\beta} - \Gamma^\beta_{\mu\kappa} g^{\mu\alpha} + \Gamma^\alpha_{\kappa\mu} g^{\mu\beta} + \Gamma^\beta_{\kappa\mu} g^{\alpha\mu} = 0. \tag{3.168}$$

Note that we have assumed a torsionless manifold, $\Gamma^\alpha_{\kappa\mu} = \Gamma^\alpha_{\mu\kappa}$.

What we derived in equation (3.162) is for the contravariant component of a rank-2 tensor. Suppose we have a rank-2 tensor, $\overleftrightarrow{t}$, for which we want to find the contravariant derivative from its components, for example in mixed form. In other words, we want to find $\nabla_\kappa t^{\alpha\beta}$ whereas what we are given is t^α_μ. In such cases, we can express the contravariant components of this tensor by applying tensor contraction. Using the metric tensor, one can write

$$t^{\alpha\beta} = g^{\beta\mu} t^\alpha_\mu, \tag{3.169}$$

so that

$$\nabla_\kappa t^{\alpha\beta} = \nabla_\kappa\left(g^{\beta\mu} t^\alpha_\mu\right) = \left(\nabla_\kappa g^{\beta\mu}\right) t^\alpha_\mu + g^{\beta\mu} \nabla_\kappa t^\alpha_\mu, \tag{3.170}$$

since from example 3.4

$$\nabla_\kappa g^{\beta\mu} = 0, \tag{3.171}$$

we can express

$$\nabla_\kappa t^{\alpha\beta} = g^{\beta\mu} \nabla_\kappa t^\alpha_\mu. \tag{3.172}$$

which the covariant derivative $t^{\alpha\beta}$ determined from the mixed component t^α_μ.

3.9 Intrinsic derivative

Like vectors (rank-1 tensors), tensor of rank-2 or higher can depend on a submanifold instead of the entire manifold. For example, a rank-2 tensor $\overleftrightarrow{t}$ can depend on a curve C on the manifold that is defined by some parameter u, $(x^\alpha(u))$. In terms of this parameter, the contravariant components of the tensor $\overleftrightarrow{t}$ can be expressed as

$$\overleftrightarrow{t}(u) = t^{\alpha\beta}\hat{e}_\alpha(u) \otimes \hat{e}_\beta(u). \tag{3.173}$$

The intrinsic derivative for this rank-2 tensor is given by

$$\frac{d\overleftrightarrow{t}(u)}{du} = \frac{d}{du}\left[t^{\alpha\beta}\hat{e}_\alpha(u) \otimes \hat{e}_\beta(u)\right]$$
$$= \frac{dt^{\alpha\beta}}{du}\hat{e}_\alpha(u) \otimes \hat{e}_\beta(u) + t^{\alpha\beta}\frac{d\hat{e}_\alpha(u)}{du} \otimes \hat{e}_\beta(u) + t^{\alpha\beta}\hat{e}_\alpha(u) \otimes \frac{d\hat{e}_\beta(u)}{du}. \tag{3.174}$$

Using the derivative of the basis vectors,

$$\frac{d\hat{e}_\alpha(u)}{du} = \frac{\partial\hat{e}_\alpha(u)}{\partial x^\kappa}\frac{dx^\kappa}{du}, \quad \frac{d\hat{e}_\beta(u)}{du} = \frac{\partial\hat{e}_\beta(u)}{\partial x^\kappa}\frac{dx^\kappa}{du}, \tag{3.175}$$

and the affine connections,

$$\frac{\partial\hat{e}_\alpha(u)}{\partial x^\kappa} = \partial_\kappa\hat{e}_\alpha = \Gamma^\eta_{\alpha\kappa}\hat{e}_\eta(u), \tag{3.176}$$

one can write

$$\frac{d\hat{e}_\alpha(u)}{du} = \Gamma^\eta_{\alpha\kappa}\frac{dx^\kappa}{du}\hat{e}_\eta(u), \quad \frac{d\hat{e}_\beta(u)}{du} = \Gamma^\eta_{\beta\kappa}\frac{dx^\beta}{du}\hat{e}_\eta(u). \tag{3.177}$$

Upon substituting equation (3.177) into equation (3.174), one finds

$$\frac{d\overleftrightarrow{t}(u)}{du} = \frac{dt^{\alpha\beta}}{du}\left[\hat{e}_\alpha(u) \otimes \hat{e}_\beta(u)\right] + t^{\alpha\beta}\Gamma^\eta_{\alpha\kappa}\frac{dx^\kappa}{du}\left[\hat{e}_\eta(u) \otimes \hat{e}_\beta(u)\right]$$
$$+ t^{\alpha\beta}\Gamma^\eta_{\beta\kappa}\frac{dx^\beta}{du}\left[\hat{e}_\alpha(u) \otimes \hat{e}_\eta(u)\right]. \tag{3.178}$$

Replacing α by η in the first term, we can write

$$\frac{dt^{\alpha\beta}}{du}\left[\hat{e}_\alpha(u) \otimes \hat{e}_\beta(u)\right] = \frac{dt^{\eta\beta}}{du}\hat{e}_\eta(u) \otimes \hat{e}_\beta(u), \tag{3.179}$$

and making the changes $\beta \to \mu$ followed by $\alpha \to \beta$ in the third term:

$$t^{\alpha\beta}\Gamma^\eta_{\beta\kappa}\frac{dx^\beta}{du}\hat{e}_\alpha(u) \otimes \hat{e}_\eta(u) = t^{\alpha\mu}\Gamma^\eta_{\mu\kappa}\frac{dx^\mu}{du}\hat{e}_\alpha(u) \otimes \hat{e}_\eta(u)$$
$$= t^{\beta\mu}\Gamma^\eta_{\mu\kappa}\frac{dx^\mu}{du}\hat{e}_\beta(u) \otimes \hat{e}_\eta(u). \tag{3.180}$$

Now, substituting equations (3.179) and (3.180) into equation (3.178), we find

$$\frac{d\overleftrightarrow{t}(u)}{du} = \frac{dt^{\eta\beta}}{du}\hat{e}_\eta(u) \otimes \hat{e}_\beta(u) + t^{\alpha\beta}\Gamma^\eta_{\alpha\kappa}\frac{dx^\kappa}{du}\left[\hat{e}_\eta(u) \otimes \hat{e}_\beta(u)\right]$$

$$+ t^{\beta\mu}\Gamma^\eta_{\mu\kappa}\frac{dx^\mu}{du}\hat{e}_\beta(u) \otimes \hat{e}_\eta(u) \qquad (3.181)$$

$$= \left[\frac{dt^{\eta\beta}}{du} + t^{\alpha\beta}\Gamma^\eta_{\alpha\kappa}\frac{dx^\kappa}{du} + t^{\beta\mu}\Gamma^\eta_{\mu\kappa}\frac{dx^\mu}{du}\right]\left[\hat{e}_\eta(u) \otimes \hat{e}_\beta(u)\right],$$

which we can put in the form

$$\frac{d\overleftrightarrow{t}(u)}{du} = \frac{Dt^{\eta\beta}}{Du}\hat{e}_\eta(u) \otimes \hat{e}_\beta(u), \qquad (3.182)$$

where

$$\frac{Dt^{\eta\beta}}{Du} = \frac{dt^{\eta\beta}}{du} + t^{\alpha\beta}\Gamma^\eta_{\alpha\kappa}\frac{dx^\kappa}{du} + t^{\beta\mu}\Gamma^\eta_{\mu\kappa}\frac{dx^\mu}{du} \qquad (3.183)$$

is called the *intrinsic (absolute) derivative of the component* $t^{\eta\beta}$ along the curve defined by $x^\alpha(u)$. For the sake of convenience we make a change of indices ($\eta \to \alpha$ and $\alpha \to \mu$) in equation (3.182) so that one can write

$$\frac{d\overleftrightarrow{t}(u)}{du} = \frac{Dt^{\alpha\beta}}{Du}\hat{e}_\alpha(u) \otimes \hat{e}_\beta(u), \qquad (3.184)$$

where

$$\frac{Dt^{\alpha\beta}}{Du} = \frac{dt^{\alpha\beta}}{du} + t^{\mu\beta}\Gamma^\alpha_{\mu\kappa}\frac{dx^\kappa}{du} + t^{\beta\mu}\Gamma^\alpha_{\mu\kappa}\frac{dx^\mu}{du}. \qquad (3.185)$$

We can switch the dummy indices κ and μ in the third term as the affine connection is symmetric for a torsionless manifold:

$$\frac{Dt^{\alpha\beta}}{Du} = \frac{dt^{\alpha\beta}}{du} + t^{\mu\beta}\Gamma^\alpha_{\mu\kappa}\frac{dx^\kappa}{du} + t^{\beta\kappa}\Gamma^\alpha_{\kappa\mu}\frac{dx^\kappa}{du}. \qquad (3.186)$$

It can be easily shown that

$$\frac{d\overleftrightarrow{t}(u)}{du} = \frac{Dt^{\alpha\beta}}{Du}\hat{e}_\alpha(u) \otimes \hat{e}_\beta(u) = \frac{Dt_{\alpha\beta}}{Du}\hat{e}^\alpha(u) \otimes \hat{e}^\beta(u)$$

$$= \frac{Dt^\alpha_\beta}{Du}\hat{e}^\alpha(u) \otimes \hat{e}_\beta(u). \qquad (3.187)$$

See problem 8.

We recall from chapter 2 that a vector (a rank-1 tensor),

$$\vec{v} = v_\mu(u)\hat{e}^\mu(u), \qquad (3.188)$$

is parallel transported when the *intrinsic derivative of this vector is zero*:

$$\frac{Dv_\mu}{Du} = \frac{dv_\mu}{du} - \Gamma^\beta_{\mu\kappa} v_\beta \frac{dx^\kappa}{du} = 0$$

$$\Rightarrow \frac{dv_\mu}{du} = \Gamma^\beta_{\mu\kappa} v_\beta \frac{dx^\kappa}{du}. \tag{3.189}$$

Like vectors, a rank-2 tensor is parallel transported when its intrinsic derivative is zero:

$$\frac{Dt^{\alpha\beta}}{Du} = 0$$

$$\Rightarrow \frac{dt^{\alpha\beta}}{du} + t^{\mu\beta}\Gamma^\alpha_{\mu\kappa}\frac{dx^\kappa}{du} + t^{\beta\kappa}\Gamma^\alpha_{\kappa\mu}\frac{dx^\kappa}{du} = 0. \tag{3.190}$$

If the tensor is defined throughout the manifold not only on a curve, then one can write

$$\frac{dt^{\alpha\beta}}{du} = \frac{\partial t^{\alpha\beta}}{\partial x^\kappa}\frac{dx^\kappa}{du} = (\partial_\kappa t^{\alpha\beta})\frac{dx^\kappa}{du}, \tag{3.191}$$

and the intrinsic derivative in equation (3.185) can be put in the form

$$\frac{Dt^{\alpha\beta}}{Du} = (\partial_\kappa t^{\alpha\beta})\frac{dx^\kappa}{du} + t^{\mu\beta}\Gamma^\alpha_{\mu\kappa}\frac{dx^\kappa}{du} + t^{\beta\kappa}\Gamma^\alpha_{\kappa\mu}\frac{dx^\kappa}{du}$$

$$\Rightarrow \frac{Dt^{\alpha\beta}}{Du} = \left(\partial_\kappa t^{\alpha\beta} + t^{\mu\beta}\Gamma^\alpha_{\mu\kappa} + t^{\beta\kappa}\Gamma^\alpha_{\kappa\mu}\right)\frac{dx^\kappa}{du}. \tag{3.192}$$

Using the result in equation (3.162), we can write

$$\frac{Dt^{\alpha\beta}}{Du} = \nabla_\kappa t^{\alpha\beta}\frac{dx^\kappa}{du}, \tag{3.193}$$

which has a much simpler form.

3.10 Homework assignment

Problem 1. Show that a second-rank tensor is linear.

Problem 2. Let us reconsider the three-dimensional sphere embedded in a four-dimensional Euclidean manifold with the covariant component of the metric tensor

$$g_{11} = \frac{a^2}{a^2 - (x^1)^2}, \; g_{22} = (x^1)^2, \; g_{33} = (x^1)^2 \sin^2(x^2),$$

$$g_{12} = g_{21} = g_{13} = g_{31} = g_{23} = g_{32} = 0.$$

Find the metric tensor components $g^{\alpha\beta}$, g^α_β, and g^β_α.

Problem 3. We recall that the metric tensor can be used to lower or raise indices. Using this property of the metric tensor for a Lorentz transformation in Cartesian coordinates for the boost, find

$$
\left[\Lambda^{\,\prime\beta}_{\mu}\right] = \begin{bmatrix} \cosh(\psi) & -\sinh(\psi) & 0 & 0 \\ -\sinh(\psi) & \cosh(\psi) & 0 & 0 \\ 0 & 0 & 1 & 0 \\ 0 & 0 & 0 & 1 \end{bmatrix}
$$

from

$$
\left[\Lambda^{\kappa}_{\alpha}\right] = \begin{bmatrix} \cosh(\psi) & \sinh(\psi) & 0 & 0 \\ \sinh(\psi) & \cosh(\psi) & 0 & 0 \\ 0 & 0 & 1 & 0 \\ 0 & 0 & 0 & 1 \end{bmatrix}.
$$

Problem 4. Show that the covariant derivatives of the mixed and covariant components of a second-rank tensor $\overleftrightarrow{t}$ are given by

$$
\nabla_{\kappa} t^{\alpha}_{\beta} = \partial_{\kappa} t^{\alpha}_{\beta} + \Gamma^{\alpha}_{\mu\kappa} t^{\mu}_{\beta} - \Gamma^{\mu}_{\beta\kappa} t^{\alpha}_{\mu},
$$
$$
\nabla_{\kappa} t_{\alpha\beta} = \partial_{\kappa} t_{\alpha\beta} - \Gamma^{\mu}_{\alpha\kappa} t_{\mu\beta} - \Gamma^{\mu}_{\beta\kappa} t_{\alpha\mu}.
$$

Using the relation

$$
\partial_{\kappa} \hat{e}^{\alpha} = -\Gamma^{\alpha}_{\beta\kappa} \hat{e}^{\beta}.
$$

Problem 5. Find the explicit form of $t_{[\alpha\beta](\kappa\mu)}$.

Problem 6. Raise the tensor $t_{\alpha\beta\kappa}$ to $t^{\alpha}_{\beta\kappa}$.

Problem 7. For the coordinate transformation $x^{\alpha} \rightarrow x^{\prime\alpha}$, show that the contravariant and mixed components of a rank-2 tensor transform as

$$
t^{\prime\alpha\beta} = t'\left(\hat{e}'_{\alpha}, \hat{e}'_{\beta}\right) = \frac{\partial x^{\prime\alpha}}{\partial x^{\kappa}} \frac{\partial x^{\prime\beta}}{\partial x^{\mu}} t^{\kappa\mu},
$$
$$
t^{\prime\beta}_{\alpha} = t'(\hat{e}'_{\alpha}, \hat{e}^{\beta'}) = \frac{\partial x^{\kappa}}{\partial x^{\prime\alpha}} \frac{\partial x^{\prime\beta}}{\partial x^{\mu}} t^{\mu}_{\kappa},
$$
$$
t^{\prime\kappa}_{\alpha\beta} = t\left(\hat{e}'_{\alpha}, \hat{e}'_{\beta}, \hat{e}^{\prime\kappa}\right) = \frac{\partial x^{\mu}}{\partial x^{\prime\alpha}} \frac{\partial x^{\nu}}{\partial x^{\prime\beta}} \frac{\partial x^{\prime\kappa}}{\partial x^{\eta}} t^{\eta}_{\mu\nu}.
$$

Problem 8. For a rank-2 tensor,

$$\overleftrightarrow{t}(u) = t^{\alpha\beta}\hat{e}_\alpha(u) \otimes \hat{e}_\beta(u) = t_{\alpha\beta}\hat{e}^\alpha(u) \otimes \hat{e}^\beta(u)$$
$$= t^\alpha_\beta \hat{e}^\alpha(u) \otimes \hat{e}_\beta(u),$$

show that

$$\frac{d\overleftrightarrow{t}(u)}{du} = \frac{Dt^{\alpha\beta}}{Du}\hat{e}_\alpha(u) \otimes \hat{e}_\beta(u) = \frac{Dt_{\alpha\beta}}{Du}\hat{e}^\alpha(u) \otimes \hat{e}^\beta(u)$$

$$= \frac{Dt^\alpha_\beta}{Du}\hat{e}^\alpha(u) \otimes \hat{e}_\beta(u).$$

(3.194)

IOP Publishing

Studies in Theoretical Physics, Volume 2
Advanced mathematical methods
Daniel Erenso

Chapter 4

Tensor application: relativistic electrodynamics

This chapter illustrates the application of tensors in relativistic electrodynamics. It begins with examining the Lorentz force and its correlation with the electromagnetic (EM) field tensor. The discussion extends to the four-current density and the continuity equation, explaining the principles underlying charge distribution and conservation. Maxwell's Equations are then introduced in connection with the EM field tensor, emphasizing their pivotal role in describing electromagnetic fields within the framework of relativity. The scalar potential, vector potential, and their relationships with the EM field tensor are explored in detail, paving the way for an in-depth understanding of electromagnetic potentials in a relativistic context. The chapter also covers gauge transformation, explicitly focusing on Maxwell's equations in the Lorentz gauge. This chapter also introduces the charged particle equation of motion, offering insights into the dynamics of charged entities within relativistic electromagnetic field tensors. At the end, the chapter lists a set of problems to solve.

4.1 The Lorentz force and the electromagnetic field tensor

Let us consider a charged particle with charge q and rest mass m_0 aboard an alien spaceship (S' frame) traveling with a velocity $\vec{u}$ as shown in figure 4.1. The spaceship is traveling in a region where there is a strong electric field, $\vec{E}$, and magnetic field, $\vec{B}$. In example 2.8, we have shown that the particle experiences a *Lorentz force (the three-force)*

$$\vec{f} = q(\vec{E} + \vec{u} \times \vec{B}), \tag{4.1}$$

which we expressed as a four-force:

$$\vec{F} = F^{\alpha}\hat{e}_{\alpha} = [F^{\alpha}] = q\gamma_u\left[\frac{(\vec{u} \cdot \vec{E})}{c}, \vec{E} + \vec{u} \times \vec{B}\right]. \tag{4.2}$$

doi:10.1088/978-0-7503-4861-4ch4

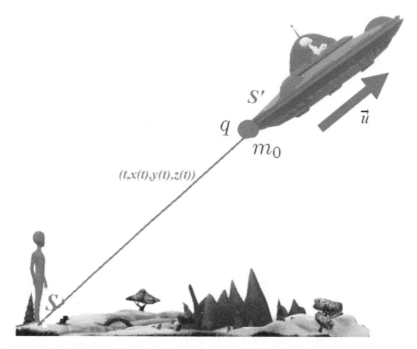

Figure 4.1. A charged particle with rest mass m_0 aboard a spaceship traveling with a velocity $\vec{u}$.

Furthermore, as we have seen, since

$$\vec{U} \cdot \vec{F} = u_\alpha F^\alpha = 0 \Rightarrow \frac{dm}{d\tau} = 0 \Rightarrow m = m_0, \tag{4.3}$$

the four-Lorentz force is a pure force that is a rank-1 tensor. A rank-1 tensor can be determined by contracting a rank-2 tensor with a rank-1 tensor. For example, we have seen that the four-wave vector of a photon in Doppler effect in the S' frame is determined from the four-wave vector in the S frame using

$$k'^u = \vec{K} \cdot \hat{e}'^\alpha = k^\beta \hat{e}_\beta \cdot \hat{e}'^\alpha = k^\beta \Lambda'^\alpha_\kappa \hat{e}_\beta \cdot \hat{e}^\kappa = \Lambda'^\alpha_\kappa \delta^\kappa_\beta k^\beta$$

$$\Rightarrow k'^u = \Lambda'^\alpha_\beta k^\beta \Rightarrow \vec{K}' = \overset{\leftrightarrow}{\Lambda}' \cdot \vec{K}, \tag{4.4}$$

where $\vec{K}'$ is a rank-1 tensor obtained by contracting the rank-2 coordinate transformation tensor, $\overset{\leftrightarrow}{\Lambda}'$ for the Minkowski space-time manifold with a rank-1 tensor, $\vec{K}$, which are given by

$$\vec{K}' = \begin{bmatrix} k'^0 \\ k'^1 \\ k'^2 \\ k'^3 \end{bmatrix}, \overset{\leftrightarrow}{\Lambda}' = \left[\Lambda'^\alpha_\beta\right] = \begin{bmatrix} \gamma & -\gamma\beta & 0 & 0 \\ -\gamma\beta & \gamma & 0 & 0 \\ 0 & 0 & 1 & 0 \\ 0 & 0 & 0 & 1 \end{bmatrix}, \vec{K} = \begin{bmatrix} k^0 \\ k^1 \\ k^2 \\ k^3 \end{bmatrix}. \tag{4.5}$$

The four-Lorentz force is a rank-1 tensor that is a function of the electric field, the magnetic field, and the velocity vectors, which all are rank-1 tensors. Suppose we define a second-rank tensor, the component of which is yet unknown, that is a function of the electric and magnetic field vectors, $\overleftrightarrow{\Omega}(\vec{E},\ \vec{B})$. Furthermore, we use the four-velocity of the charged particle, $\vec{U}$. Then one can express the four-Lorentz force in equation (4.2) in terms of the second-rank electromagnetic (EM) field tensor, $\overleftrightarrow{\Omega}$, contracted by the first-rank four-velocity tensor, $\vec{U}$, as

$$\vec{F} = q\left(\overleftrightarrow{\Omega}\cdot\vec{U}\right). \tag{4.6}$$

We do know the four-velocity components of the particle, but we do not know the 16 components of the EM field tensor. We must determine these components in terms of the electric and magnetic field vectors components. To this end, let us express the EM field tensor $\overleftrightarrow{\Omega}$ in terms of its covariant components and the four-velocity $\vec{U}$ in terms of its contravariant components, so that the four-force can be expressed as

$$\vec{F} = F^\alpha \hat{e}_\alpha = [F_\alpha] = \left[q\Omega_{\alpha\beta}u^\beta\right]. \tag{4.7}$$

As we have shown in example 2.8, the four-Lorentz force is a pure force and one must have

$$\vec{F}\cdot\vec{U} = 0 \Rightarrow F_\alpha U^\alpha = q\Omega_{\alpha\beta}U^\beta U^\alpha = 0,$$
$$\vec{U}\cdot\vec{F} = 0 \Rightarrow U^\beta F_\beta = qU^\beta\Omega_{\beta\alpha}U^\alpha = q\Omega_{\beta\alpha}U^\alpha U^\beta = 0, \tag{4.8}$$

so that, upon adding these two equations, one finds

$$\left(\Omega_{\alpha\beta} + \Omega_{\beta\alpha}\right)U^\beta U^\alpha = 0. \tag{4.9}$$

There follows that

$$\Omega_{\alpha\beta} = -\Omega_{\beta\alpha}, \tag{4.10}$$

which means the EM field tensor is an *antisymmetric* tensor. Noting that

$$U^\beta = g^{\beta\kappa}U_\kappa,\ U^\alpha = g^{\alpha\mu}U_\mu, \tag{4.11}$$

Equation (4.9) can be rewritten:

$$\left(\Omega_{\alpha\beta} + \Omega_{\beta\alpha}\right)g^{\beta\kappa}U_\kappa g^{\alpha\mu}U_\mu = 0$$
$$\Rightarrow \left(g^{\alpha\mu}g^{\beta\kappa}\Omega_{\alpha\beta} + g^{\beta\kappa}g^{\alpha\mu}\Omega_{\beta\alpha}\right)U_\kappa U_\mu = 0, \tag{4.12}$$

so that using

$$g^{\alpha\mu}g^{\beta\kappa}\Omega_{\alpha\beta} = \Omega^{\mu\kappa},\ g^{\beta\kappa}g^{\alpha\mu}\Omega_{\beta\alpha} = \Omega^{\kappa\mu},$$

equation (4.12) becomes

$$(\Omega^{\mu\kappa} + \Omega^{\kappa\mu})U_\kappa U_\mu = 0 \Rightarrow \Omega^{\mu\kappa} = -\Omega^{\kappa\mu}. \tag{4.13}$$

In view of the result in equation (4.13), one finds

$$\Omega^{10} = -\Omega^{01},\ \Omega^{20} = -\Omega^{02},\ \Omega^{30} = -\Omega^{03},$$
$$\Omega^{21} = -\Omega^{12},\ \Omega^{31} = -\Omega^{13},\ \Omega^{32} = -\Omega^{23},$$
$$\Omega^{\mu\mu} = -\Omega^{\mu\mu} \Rightarrow 2\Omega^{\mu\mu} = 0 \Rightarrow \Omega^{\mu\mu} = 0$$
$$\Rightarrow \Omega^{00} = \Omega^{11} = \Omega^{22} = \Omega^{33} = 0, \tag{4.14}$$

and for the contravariant components of the EM field tensor,

$$\overleftrightarrow{\Omega} = [\Omega^{\alpha\beta}] = \begin{Bmatrix} \Omega^{00} & \Omega^{01} & \Omega^{02} & \Omega^{03} \\ \Omega^{10} & \Omega^{11} & \Omega^{12} & \Omega^{13} \\ \Omega^{20} & \Omega^{21} & \Omega^{22} & \Omega^{23} \\ \Omega^{30} & \Omega^{31} & \Omega^{32} & \Omega^{33} \end{Bmatrix}, \tag{4.15}$$

one can write

$$\overleftrightarrow{\Omega} = [\Omega^{\alpha\beta}] = \begin{Bmatrix} 0 & \Omega^{01} & \Omega^{02} & \Omega^{03} \\ -\Omega^{01} & 0 & \Omega^{12} & \Omega^{13} \\ -\Omega^{02} & -\Omega^{12} & 0 & \Omega^{23} \\ -\Omega^{03} & -\Omega^{13} & -\Omega^{23} & 0 \end{Bmatrix}. \tag{4.16}$$

The result in equation (4.16) shows that there are still six independent components that we need to determine for the EM field tensor. We will determine these components after we have introduced the four-current density and Maxwell's equations.

4.2 The four-current density and the continuity equation

In electrodynamics what we know is the charge density ρ, a scalar (zero-rank tensor), and the volume current density $\vec{j}$, a rank-1 tensor. Next we will see how these two quantities combine to form the four-current density. To this end, let us consider a cube with a proper side length l_0 that has N particles with charge q each on the S' frame as shown in figure 4.2. The particles are uniformly distributed in the volume. The 'proper number density' for the charges in this cube n_0 can then be written as

$$n_0 = \frac{N}{l_0^3}. \tag{4.17}$$

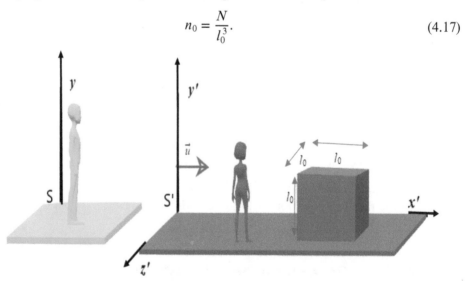

Figure 4.2. A cube with length l_0 and uniform charge density on an inertial frame S' moving with a velocity u in the positive x-direction. The cube has N point charges uniformly distributed in the volume $V_0 = l_0^3$.

Note that the proper length and the proper number density are measured by an observer at rest. In this case, the proper length is the length measured by the woman in the S' frame as shown in figure 4.2. Then for the 'proper charge density' ρ_0 and current density $\vec{j}\,'$, one can write

$$\rho_0 = qn_0, \vec{j}\,' = 0, \tag{4.18}$$

since each particle carries a charge q. Let the S' frame where the charges in the cube are sitting at rest move with a velocity $\vec{u}$ along the positive x-direction as measured by an observer in the S frame (the man as shown in figure 4.2). For an observer in the S frame, since the length of the cube along the x-direction is Lorentz contracted,

$$l = l_0\sqrt{1 - \frac{u^2}{c^2}} = \frac{l_0}{\gamma_u}, \tag{4.19}$$

where

$$\gamma_u = \frac{1}{\sqrt{1 - \frac{u^2}{c^2}}}, \tag{4.20}$$

the charge density becomes

$$\rho = qn = q\frac{N}{ll_0^2} = \gamma_u q\frac{N}{l_0^3} = \gamma_u \rho_0, \tag{4.21}$$

where we replaced $\rho_0 = N/l_0^3$. Since each charge inside the cube with charge density ρ is also moving with a velocity $\vec{u}$ in the positive x-direction, there is a current for an observer in the S frame. The current density can be expressed as

$$\vec{j} = \rho\vec{u} = \gamma_u \rho_0 \vec{u}. \tag{4.22}$$

Therefore, the charge and current densities in the S frame become

$$\rho = \gamma_u \rho_0, \vec{j} = \gamma_u \rho_0 \vec{u}. \tag{4.23}$$

From equations (4.18) and (4.23), we note that the transformation for the charge and current densities from the S' to S frames is

$$\left(\rho_0, \vec{j}\,' = 0\right) \rightarrow (\rho, \vec{j}\,). \tag{4.24}$$

This suggests that the three-current can be replaced by the four-current. We define the zero-component of the four-current density in the S' frame and S frame by

$$j'^0 = c\rho_0, j^0 = c\rho, \tag{4.25}$$

respectively. Then the four-current densities become

$$\vec{J} = j^\alpha \hat{e}_\alpha = (j^0, j^1, j^2, j^3) = \left(c\rho, j_x, j_y, j_z\right) = (c\rho, \vec{j}\,), \tag{4.26}$$

$$\vec{J}' = j'^\alpha \hat{e}'_\alpha = (j'^0, j'^1, j'^2, j'^3) = \left(c\rho_0, 0\right). \tag{4.27}$$

Recalling that tensors are geometrical properties of the manifold, they must remain the same in any reference frame. Therefore, noting that the inner product for the current density in the S' frame,

$$\vec{J}' \cdot \vec{J}' = (c\rho_0)^2, \tag{4.28}$$

we must also find in the S frame

$$\vec{J} \cdot \vec{J} = \vec{J}' \cdot \vec{J}' = (c\rho_0)^2. \tag{4.29}$$

In terms of the four-velocity, the contravariant components of the four-current density in the S frame can be expressed as

$$\vec{J} = j^\alpha \hat{e}_\alpha = [j^\alpha] = (c\rho, \vec{j}) = \left(c\gamma_u \rho_0, \gamma_u \rho_0 \vec{u} \right) = \gamma_u \rho_0 (c, \vec{u}). \tag{4.30}$$

Note that ρ is the charge density of a volume of charge moving with a velocity $\vec{u}$ as measured by an observer in the S frame.

In electrodynamics we have seen that the charge density, ρ, and the three-current density, $\vec{j}$, are related by the continuity equation:

$$\frac{\partial \rho}{\partial t} + \nabla \cdot \vec{j} = 0, \tag{4.31}$$

which is conservation of charges. Next we will see that the continuity equation can be expressed in terms of the four-current density in a very simple way. To this end, using

$$\frac{\partial \rho}{\partial t} = \frac{\partial (c\rho)}{\partial (ct)} = \frac{\partial}{\partial x^0} j^0 = \partial_0 j^0, \tag{4.32}$$

$$\nabla \cdot \vec{j} = \frac{\partial j_x}{\partial x} + \frac{\partial j_y}{\partial y} + \frac{\partial j_z}{\partial z} = \frac{\partial}{\partial x^1} j^1 + \frac{\partial}{\partial x^2} j^2 + \frac{\partial}{\partial x^3} j^3$$
$$\Rightarrow \nabla \cdot \vec{j} = \partial_1 j^1 + \partial_2 j^2 + \partial_3 j^3, \tag{4.33}$$

one can rewrite

$$\frac{\partial \rho}{\partial t} + \nabla \cdot \vec{j} = \partial_0 j^0 + \partial_1 j^1 + \partial_2 j^2 + \partial_3 j^3 = 0$$
$$\Rightarrow \partial_\alpha j^\alpha = \vec{\nabla} \cdot \vec{J} = 0, \tag{4.34}$$

where $\vec{\nabla}$ is the four-gradient,

$$\vec{\nabla} = \left[\frac{\partial}{\partial (ct)}, \frac{\partial}{\partial x}, \frac{\partial}{\partial y}, \frac{\partial}{\partial z} \right] = \left[\frac{\partial}{\partial x^0}, \frac{\partial}{\partial x^1}, \frac{\partial}{\partial x^2}, \frac{\partial}{\partial x^3} \right] \tag{4.35}$$
$$\Rightarrow \vec{\nabla} = \hat{e}^\alpha \partial_\alpha = [\partial_0, \partial_1, \partial_2, \partial_3],$$

and $\vec{J}$ is the four-current density,

$$\vec{J} = \left(c\rho, j_x, j_y, j_z\right) = (j^0, j^1, j^2, j^3) = j^\alpha \hat{e}_\alpha, \tag{4.36}$$

in the Minkowski space-time manifold. Equation (4.34) is the continuity equation.

4.3 Maxwell's equations and the electromagnetic field tensor

From electromagnetism, we know that the electric field, $\vec{E}$, and magnetic field, $\vec{B}$, are determined by Maxwell's equation:

$$\nabla \cdot \vec{E} = \frac{\rho}{\epsilon_0} = \mu_0 \frac{c^2\rho}{\epsilon_0}, \quad \nabla \times \vec{B} - \epsilon_0\mu_0 \frac{\partial \vec{E}}{\partial t} = \mu_0 \vec{j}, \tag{4.37}$$

$$\nabla \times \vec{E} + \frac{\partial \vec{B}}{\partial t} = 0, \quad \nabla \cdot \vec{B} = 0. \tag{4.38}$$

Note that the electrical permittivity, ϵ_0, and the magnetic permeability, μ_0, of a free space are universal constants. These equations obviously indicate that the electric or magnetic fields are a result of some charge density ρ and three-current density $\vec{j}$, which can be both described as a four-current density $\vec{J}$ in the Minkowski space-time manifold as we saw in the previous section. In fact, we can see that using $c^2 = 1/\epsilon_0\mu_0$, by rewriting equations (4.37)–(4.31) this can be written as

$$\nabla \cdot \vec{E} = \nabla \cdot \left(\frac{\vec{E}}{c}\right) = \mu_0(c\rho) = \mu_0 j^0, \tag{4.39}$$

$$\nabla \times \vec{B} - \epsilon_0\mu_0 \frac{\partial \vec{E}}{\partial t} = \nabla \times \vec{B} - \frac{\partial}{\partial(ct)}\left(\frac{\vec{E}}{c}\right) = \mu_0 \vec{j}, \tag{4.40}$$

$$\nabla \times \vec{E} + \frac{\partial \vec{B}}{\partial t} = \nabla \times \left(\frac{\vec{E}}{c}\right) + \frac{\partial \vec{B}}{\partial(ct)} = 0, \tag{4.41}$$

$$\nabla \cdot \vec{B} = 0. \tag{4.42}$$

Applying the relations in equation (4.35) and noting that

$$\vec{B} = \left[B_x, B_y, B_z\right] = [B^1, B^2, B^3],$$
$$\frac{\vec{E}}{c} = \left[\frac{E_x}{c}, \frac{E_y}{c}, \frac{E_z}{c}\right] = \left[\frac{E^1}{c}, \frac{E^2}{c}, \frac{E^3}{c}\right],$$

one can rewrite equations (4.39)–(4.42) as

$$\frac{1}{c}(\partial_1 E^1 + \partial_2 E^2 + \partial_3 E^3) = \mu_0 j^0, \tag{4.43}$$

$$\left(\partial_2 B^3 - \partial_3 B^2\right)\hat{e}_1 + \left(\partial_3 B^1 - \partial_1 B^3\right)\hat{e}_2 + \left(\partial_1 B^2 - \partial_2 B^1\right)\hat{e}_3$$

$$-\frac{1}{c}\left[\left(\partial_0 E^1\right)\hat{e}_1 + \left(\partial_0 E^2\right)\hat{e}_2 + \left(\partial_0 E^3\right)\hat{e}_3\right] = \mu_0 \vec{j},$$

$$\Rightarrow \left(\partial_2 B^3 - \partial_3 B^2 - \frac{1}{c}\partial_0 E^1\right)\hat{e}_1 + \left(\partial_3 B^1 - \partial_1 B^3 - \frac{1}{c}\partial_0 E^2\right)\hat{e}_2 \tag{4.44}$$

$$+ \left(\partial_1 B^2 - \partial_2 B^1 - \frac{1}{c}\partial_0 E^3\right)\hat{e}_3 = \mu_0 \vec{j},$$

$$\frac{1}{c}\left[\left(\partial_2 E^3 - \partial_3 E^2\right)\hat{e}_1 + \left(\partial_3 E^1 - \partial_1 E^3\right)\hat{e}_2 + \left(\partial_1 E^2 - \partial_2 E^1\right)\hat{e}_3\right]$$

$$+ \left(\partial_0 B^1\right)\hat{e}_1 + \left(\partial_0 B^2\right)\hat{e}_2 + \left(\partial_0 B^3\right)\hat{e}_3$$

$$\Rightarrow \left(\partial_0 B^1 + \frac{1}{c}\left(\partial_2 E^3 - \partial_3 E^2\right)\right)\hat{e}_1 + \left(\partial_0 B^2 + \frac{1}{c}\left(\partial_3 E^1 - \partial_1 E^3\right)\right)\hat{e}_2 \tag{4.45}$$

$$+ \left(\partial_0 B^3 \frac{1}{c}\left(\partial_1 E^2 - \partial_2 E^1\right)\right)\hat{e}_3 = 0,$$

and

$$\partial_1 B^1 + \partial_2 B^2 + \partial_3 B^3 = 0. \tag{4.46}$$

These forms of Maxwell's equations indicate each one of them involves at least the three-gradient ∇ operation on the fields (on the left side), which results in μ_0 times the components of the four-current density $\mu_0[j^0, \vec{j}\,]$ or zero (on the right side). In particular, equations (4.43) and (4.44) in fact involve the four-gradient and the four-current density. Therefore, it is reasonable and well justified that Maxwell's equation can be determined by contraction of a rank-2 EM field tensor, $\overleftrightarrow{\Omega}$, by the four-gradient $\vec{\nabla}$ such that

$$\vec{\nabla} \cdot \overleftrightarrow{\Omega} = \mu_0 \vec{J}. \tag{4.47}$$

We have shown that the EM field tensor $\overleftrightarrow{\Omega}$ is an antisymmetric rank-2 tensor, the components of which are not fully determined. If the field is expressed in terms of its contravariant components, we may write equation (4.47) as

$$\partial_\alpha \Omega^{\alpha\beta} = \mu_0 j^\beta, \tag{4.48}$$

or

$$\partial_\beta \Omega^{\beta\alpha} = \mu_0 j^\alpha. \tag{4.49}$$

Taking ∂_β of equation (4.48) and ∂_α of equation (4.49), we have

$$\partial_\beta \partial_\alpha \Omega^{\alpha\beta} = \mu_0 \partial_\beta j^\beta, \tag{4.50}$$

and

$$\partial_\alpha \partial_\beta \Omega^{\beta\alpha} = \mu_0 \partial_\alpha j^\alpha, \tag{4.51}$$

so that upon adding these two equations, one finds

$$\mu_0\left(\partial_\beta j^\beta + \partial_\alpha j^\alpha\right) = \partial_\beta \partial_\alpha \Omega^{\alpha\beta} + \partial_\alpha \partial_\beta \Omega^{\beta\alpha} = \partial_\alpha \partial_\beta (\Omega^{\alpha\beta} + \Omega^{\beta\alpha})$$

$$\Rightarrow \mu_0\left(\partial_\alpha j^\alpha + \partial_\beta j^\beta\right) = \partial_\alpha \partial_\beta (\Omega^{\alpha\beta} + \Omega^{\beta\alpha}) \tag{4.52}$$

$$\Rightarrow 2\mu_0(\partial_\alpha j^\alpha) = \partial_\alpha \partial_\beta (\Omega^{\alpha\beta} + \Omega^{\beta\alpha}),$$

where we replaced the dummy index β by α in the left side of this equation. Since the EM field tensor is *antisymmetric*,

$$\Omega^{\beta\alpha} = -\Omega^{\alpha\beta}, \tag{4.53}$$

upon substituting this into equation (4.52), one finds

$$2\mu_0(\partial_\alpha j^\alpha) = 0 \Rightarrow \partial_\alpha j^\alpha = 0 \Rightarrow \frac{\partial \rho}{\partial t} + \nabla \cdot \vec{j} = 0, \tag{4.54}$$

which is *the continuity equation* in equation (4.34).

Example 4.1. Show the following in the Minkowski space-time manifold.

(a) The contravariant component for the four-gradient operator $[\partial^\alpha]$, is given by

$$[\partial^\alpha] = \begin{bmatrix} \partial_0 \\ -\partial_1 \\ -\partial_2 \\ -\partial_3 \end{bmatrix}. \tag{4.55}$$

(b) The inner product $\partial_\alpha \partial^\alpha$ is given by

$$\Box = \partial_\alpha \partial^\alpha = \frac{\partial^2}{\partial(ct)^2} - \frac{\partial^2}{\partial x^2} - \frac{\partial^2}{\partial y^2} - \frac{\partial^2}{\partial z^2}.$$

Solution:

(a) Using the metric tensor for the Minkowski space-time manifold, we can write

$$[\partial_\alpha] = \left(\frac{1}{c}\frac{\partial}{\partial t}, \frac{\partial}{\partial x}, \frac{\partial}{\partial y}, \frac{\partial}{\partial z} \right) \Rightarrow [\partial^\alpha] = \left[g^{\alpha\beta} \partial_\beta \right]$$

$$= \begin{bmatrix} 1 & 0 & 0 & 0 \\ 0 & -1 & 0 & 0 \\ 0 & 0 & -1 & 0 \\ 0 & 0 & 0 & -1 \end{bmatrix} \begin{bmatrix} \dfrac{\partial}{\partial(ct)} \\ \dfrac{\partial}{\partial x} \\ \dfrac{\partial}{\partial y} \\ \dfrac{\partial}{\partial z} \end{bmatrix}.$$

Multiplying these two matrices, one finds

$$[\partial^\alpha] = \begin{bmatrix} \partial^0 \\ \partial^1 \\ \partial^2 \\ \partial^3 \end{bmatrix} = \begin{bmatrix} \dfrac{\partial}{\partial(ct)} \\[2mm] -\dfrac{\partial}{\partial x} \\[2mm] -\dfrac{\partial}{\partial y} \\[2mm] -\dfrac{\partial}{\partial z} \end{bmatrix} = \begin{bmatrix} \partial_0 \\ -\partial_1 \\ -\partial_2 \\ -\partial_3 \end{bmatrix}. \tag{4.56}$$

(b) Using the result in part (a), one finds

$$\Box = \partial^\alpha \partial_\alpha = \left(\frac{1}{c} \frac{\partial}{\partial t}, \frac{\partial}{\partial x}, \frac{\partial}{\partial y}, \frac{\partial}{\partial z} \right) \begin{pmatrix} \dfrac{1}{c}\dfrac{\partial}{\partial t} \\[2mm] -\dfrac{\partial}{\partial x} \\[2mm] -\dfrac{\partial}{\partial y} \\[2mm] -\dfrac{\partial}{\partial z} \end{pmatrix} \tag{4.57}$$

$$= \frac{1}{c^2} \frac{\partial^2}{\partial t^2} - \frac{\partial^2}{\partial x^2} - \frac{\partial^2}{\partial y^2} - \frac{\partial^2}{\partial z^2},$$

which is known as the *d'Alembert* operator.

4.4 The scalar, the vector potentials, and the electromagnetic field tensor

Next we shall determine the EM field tensor components. To this end, we recall the scalar, V, and the vector potential, $\vec{A}$, and the relation of these potentials to the electric and magnetic fields:

$$\vec{E} = -\frac{\partial \vec{A}}{\partial t} - \nabla V, \quad \vec{B} = \nabla \times \vec{A}. \tag{4.58}$$

We recall from chapter 2 the components for the curl of a vector are given by

$$\left(\nabla \times \vec{A} \right)_{\alpha\beta} = \frac{\partial A_\beta}{\partial x^\alpha} - \frac{\partial A_\alpha}{\partial x^\beta} = \partial_\alpha A_\beta - \partial_\beta A_\alpha, \tag{4.59}$$

which forms the covariant components for *a rank-2 antisymmetric tensor*. Using

$$\vec{A} = \left[A_x, A_y, A_z\right] = A^1 \hat{e}_1 + A^2 \hat{e}_2 + A^3 \hat{e}_3 = [A^1, A^2, A^3],$$

$$\vec{B} = \left[B_x, B_y, B_z\right] = B^1 \hat{e}_1 + B^2 \hat{e}_2 + B^3 \hat{e}_3 = [B^1, B^2, B^3], \tag{4.60}$$

$$\nabla = \frac{\partial}{\partial x}\hat{x} + \frac{\partial}{\partial y}\hat{y} + \frac{\partial}{\partial z}\hat{z} = \frac{\partial}{\partial x^1}\hat{x} + \frac{\partial}{\partial x^2}\hat{y} + \frac{\partial}{\partial x^3}\hat{z},$$

for the magnetic field in equation (4.58), we can write

$$\vec{B} = \nabla \times \vec{A}$$
$$= (\partial_2 A^3 - \partial_3 A^2)\hat{x} + (\partial_3 A^1 - \partial_1 A^3)\hat{y} + (\partial_1 A^2 - \partial_2 A^1)\hat{z} \tag{4.61}$$
$$= -(\partial_2 A_3 + \partial_3 A_2)\hat{x} - (\partial_3 A_1 - \partial_1 A_3)\hat{y} + (\partial_1 A_2 - \partial_2 A_1)\hat{z}$$

$$\Rightarrow \begin{bmatrix} B^1 \\ B^2 \\ B^3 \end{bmatrix} = \begin{bmatrix} \partial_3 A_2 - \partial_2 A_3 \\ \partial_1 A_3 - \partial_3 A_1 \\ \partial_2 A_1 - \partial_1 A_2 \end{bmatrix}, \tag{4.62}$$

where we used

$$A_\alpha = g_{\alpha\beta}A^\beta \Rightarrow [A^1, A^2, A^3] = -[A_1, A_2, A_3].$$

Noting that the result in equation (4.62) clearly shows that the components of the magnetic field are the covariant components of an antisymmetric rank-2 tensor, $\Omega_{\alpha\beta}(\hat{e}^\alpha, \hat{e}^\beta)$,

$$\Omega_{\alpha\beta}(\hat{e}^\alpha, \hat{e}^\beta) = \partial_\alpha A_\beta - \partial_\beta A_\alpha \Rightarrow \Omega_{\beta\alpha}(\hat{e}^\alpha, \hat{e}^\beta) = \partial_\beta A_\alpha - \partial_\alpha A_\beta$$
$$\Rightarrow \Omega_{\beta\alpha}(\hat{e}^\alpha, \hat{e}^\beta) = -\left(\partial_\alpha A_\beta - \partial_\beta A_\alpha\right) = -\Omega_{\alpha\beta}(\hat{e}^\alpha, \hat{e}^\beta). \tag{4.63}$$

Therefore, from equation (4.62), we can establish the relation

$$\begin{bmatrix} B^1 \\ B^2 \\ B^3 \end{bmatrix} = \begin{bmatrix} \partial_3 A_2 - \partial_2 A_3 \\ \partial_1 A_3 - \partial_3 A_1 \\ \partial_2 A_1 - \partial_1 A_2 \end{bmatrix} = \begin{bmatrix} \Omega_{32}(\hat{e}^\alpha, \hat{e}^\beta) \\ \Omega_{13}(\hat{e}^\alpha, \hat{e}^\beta) \\ \Omega_{21}(\hat{e}^\alpha, \hat{e}^\beta) \end{bmatrix}. \tag{4.64}$$

Similarly, for the electric field in equation (4.58), using

$$\nabla = [\partial_1, \partial_2, \partial_3], \ \vec{A} = [A^1, A^2, A^3], \ \vec{E} = [E^1, E^2, E^3], \tag{4.65}$$

and recalling that

$$\partial^0 = \partial_0 = \frac{\partial}{\partial(ct)} = \frac{\partial}{\partial x^0}, \tag{4.66}$$

one can rewrite the electric field in equation (4.58) as

$$\frac{\vec{E}}{c} = -\frac{\partial \vec{A}}{\partial (ct)} - \nabla\left(\frac{V}{c}\right)$$

$$-\frac{\vec{E}}{c} = \partial_0(A^1\hat{e}_1 + A^2\hat{e}_2 + A^3\hat{e}_3) + (\partial_1\hat{e}^1 + \partial_2\hat{e}^2 + \partial_3\hat{e}^3)\frac{V}{c}$$

$$= \left(\partial_0 A^1\hat{e}_1 + \partial_1\left(\frac{V}{c}\right)\hat{e}^1\right) + \left(\partial_0 A^2\hat{e}_2 + \partial_2\left(\frac{V}{c}\right)\hat{e}^3\right)$$

$$+ \left(\partial_0 A^3\hat{e}_3 + \partial_3\left(\frac{V}{c}\right)\hat{e}^3\right). \tag{4.67}$$

Introducing the zeroth component:

$$A^0\hat{e}_0 = A_0\hat{e}^0 = \frac{V}{c}\hat{e}^0 = \frac{V}{c}\hat{e}_0, \tag{4.68}$$

we define the four-vector potential:

$$\vec{A} = [A^0, A^1, A^2, A^3] = \left[\frac{V}{c}, A_x, A_y, A_z\right]. \tag{4.69}$$

Then substituting equation (4.68) into equation (4.67), we can write

$$-\frac{\vec{E}}{c} = (\partial_0 A^1\hat{e}_1 + \partial_1 A_0\hat{e}^1) + (\partial_0 A^2\hat{e}_2 + \partial_2 A_0\hat{e}^3)$$

$$+ (\partial_0 A^3\hat{e}_3 + \partial_3 A_0\hat{e}^3). \tag{4.70}$$

From the result in equation (4.56), we have

$$\begin{bmatrix} \partial_1 \\ \partial_2 \\ \partial_3 \end{bmatrix} = \begin{bmatrix} -\partial^1 \\ -\partial^2 \\ -\partial^3 \end{bmatrix}, \tag{4.71}$$

and equation (4.70) can be rewritten as

$$-\frac{\vec{E}}{c} = (\partial_0 A^1 - \partial^1 A_0)\hat{e}_1 + (\partial_0 A^2 - \partial^2 A_0)\hat{e}_2$$

$$+ (\partial_0 A^3 - \partial^3 A_0)\hat{e}_3, \tag{4.72}$$

which we can put in the form

$$\begin{bmatrix} -\dfrac{E^1}{c} \\ -\dfrac{E^2}{c} \\ -\dfrac{E^3}{c} \end{bmatrix} = \begin{bmatrix} \partial_0 A^1 - \partial^1 A_0 \\ \partial_0 A^2 - \partial^2 A_0 \\ \partial_0 A^3 - \partial^3 A_0 \end{bmatrix}. \tag{4.73}$$

Noting that $\partial^\alpha \to \partial^\alpha(\hat{e}_\alpha)$, $\partial_\alpha \to \partial_\alpha(\hat{e}^\alpha)$, $A_\beta \to A_\beta(\hat{e}^\beta)$, and $A^\beta \to A^\beta(\hat{e}_\beta)$, the result in equation (4.73) clearly shows that the components of the electric field are the mixed components of an antisymmetric rank-2 tensor $\Omega_\alpha^\beta(\hat{e}_\alpha, \hat{e}_\beta)$ and one can establish the relation

$$\begin{bmatrix} -\dfrac{E^1}{c} \\[2mm] -\dfrac{E^2}{c} \\[2mm] -\dfrac{E^3}{c} \end{bmatrix} = \begin{bmatrix} \partial^0 A^1 - \partial^1 A^0 \\ \partial^0 A^2 - \partial^2 A^0 \\ \partial^0 A^3 - \partial^3 A^0 \end{bmatrix} = \begin{bmatrix} \Omega_0^1(\hat{e}^\alpha, \hat{e}_\beta) \\ \Omega_0^2(\hat{e}^\alpha, \hat{e}_\beta) \\ \Omega_0^3(\hat{e}^\alpha, \hat{e}_\beta) \end{bmatrix}. \tag{4.74}$$

Noting that

$$\Omega_{\alpha\beta} = g_{\beta\mu}\Omega_\alpha^\mu, \tag{4.75}$$

using the corresponding mixed components in equation (4.74) and the metric tensor for a Minkowski space-time manifold, we find for the covariant components:

$$\Omega_{01}(\hat{e}_\alpha, \hat{e}_\beta) = g_{1\mu}\Omega_0^\mu = g_{10}\Omega_0^0 + g_{11}\Omega_0^1 + g_{12}\Omega_0^2 + g_{13}\Omega_0^3$$

$$\Rightarrow \Omega_{01}(\hat{e}_\alpha, \hat{e}_\beta) = g_{11}\Omega_0^1 = -\Omega_0^1 = \frac{E^1}{c},$$

$$\Omega_{02}(\hat{e}_\alpha, \hat{e}_\beta) = g_{2\mu}\Omega_0^\mu = g_{20}\Omega_0^0 + g_{21}\Omega_0^1 + g_{22}\Omega_0^2 + g_{23}\Omega_0^3$$

$$\Rightarrow \Omega_{02}(\hat{e}_\alpha, \hat{e}_\beta) = g_{22}\Omega_0^2 = -\Omega_0^2 = \frac{E^2}{c},$$

$$\Omega_{03}(\hat{e}_\alpha, \hat{e}_\beta) = g_{3\mu}\Omega_0^\mu = g_{30}\Omega_0^0 + g_{31}\Omega_0^1 + g_{32}\Omega_0^2 + g_{33}\Omega_0^3$$

$$\Rightarrow \Omega_{03}(\hat{e}_\alpha, \hat{e}_\beta) = g_{33}\Omega_0^3 = -\Omega_0^3 = \frac{E^3}{c}.$$

Therefore, for the covariant component of the EM field tensor

$$\overset{\leftrightarrow}{\Omega}(\hat{e}_\alpha, \hat{e}_\beta) = \left[\Omega(\hat{e}_\alpha, \hat{e}_\beta)\right] = \begin{Bmatrix} 0 & \Omega_{01} & \Omega_{02} & \Omega_{03} \\ -\Omega_{01} & 0 & \Omega_{12} & \Omega_{13} \\ -\Omega_{02} & -\Omega_{12} & 0 & \Omega_{23} \\ -\Omega_{03} & -\Omega_{13} & -\Omega_{23} & 0 \end{Bmatrix}, \tag{4.76}$$

using the results

$$\Omega_{32}(\hat{e}^\alpha, \hat{e}^\beta) = B^1, \ \Omega_{13}(\hat{e}^\alpha, \hat{e}^\beta) = B^2, \ \Omega_{21}(\hat{e}^\alpha, \hat{e}^\beta) = B^3,$$

$$\Omega_{01}(\hat{e}_\alpha, \hat{e}_\beta) = \frac{E^1}{c}, \ \Omega_{02}(\hat{e}_\alpha, \hat{e}_\beta) = \frac{E^2}{c}, \ \Omega_{03}(\hat{e}_\alpha, \hat{e}_\beta) = \frac{E^3}{c}, \tag{4.77}$$

we find

$$\overset{\leftrightarrow}{\Omega}(\hat{e}^{\alpha}, \hat{e}^{\beta}) = \left[\Omega_{\alpha\beta}\right] = \begin{Bmatrix} 0 & \dfrac{E^1}{c} & \dfrac{E^2}{c} & \dfrac{E^3}{c} \\ -\dfrac{E^1}{c} & 0 & -B^3 & B^2 \\ -\dfrac{E^2}{c} & B^3 & 0 & -B^1 \\ -\dfrac{E^3}{c} & -B^2 & B^1 & 0 \end{Bmatrix}. \tag{4.78}$$

In terms of the electric and magnetic fields in Cartesian coordinates, equation (4.78) can be rewritten as

$$\overset{\leftrightarrow}{\Omega}(\hat{e}^{\alpha}, \hat{e}^{\beta}) = \left[\Omega_{\alpha\beta}\right] = \begin{Bmatrix} 0 & \dfrac{E_x}{c} & \dfrac{E_y}{c} & \dfrac{E_z}{c} \\ -\dfrac{E_x}{c} & 0 & -B_z & B_y \\ -\dfrac{E_y}{c} & B_z & 0 & -B_x \\ -\dfrac{E_z}{c} & -B_y & B_x & 0 \end{Bmatrix}.$$

These are the covariant components for the EM field tensor.

Example 4.2. Using the metric tensor for the Minkowski space-time manifold, find the mixed and the contravariant components of the EM field tensor.

Solution: Using the metric tensor, we can write for the mixed component

$$\left[\Omega^{\alpha}_{\beta}\right] = \left[g^{\alpha\kappa}\Omega_{\kappa\beta}\right] = \begin{pmatrix} 1 & 0 & 0 & 0 \\ 0 & -1 & 0 & 0 \\ 0 & 0 & -1 & 0 \\ 0 & 0 & 0 & -1 \end{pmatrix} \begin{pmatrix} 0 & \dfrac{E_x}{c} & \dfrac{E_y}{c} & \dfrac{E_z}{c} \\ -\dfrac{E_x}{c} & 0 & -B_z & B_y \\ -\dfrac{E_y}{c} & B_z & 0 & -B_x \\ -\dfrac{E_z}{c} & -B_y & B_x & 0 \end{pmatrix}$$

$$\Rightarrow \left[\Omega^{\beta}_{\alpha}\right] = \begin{pmatrix} 0 & \dfrac{E_x}{c} & \dfrac{E_y}{c} & \dfrac{E_z}{c} \\ \dfrac{E_x}{c} & 0 & B_z & -B_y \\ \dfrac{E_y}{c} & -B_z & 0 & B_x \\ \dfrac{E_z}{c} & B_y & -B_x & 0 \end{pmatrix}. \tag{4.79}$$

For the contravariant component, we have

$$[\Omega^{\alpha\beta}] = [g^{\kappa\beta}\Omega^{\alpha}_{\kappa}] = [\Omega][g], \tag{4.80}$$

$$= \begin{pmatrix} 0 & \dfrac{E_x}{c} & \dfrac{E_y}{c} & \dfrac{E_z}{c} \\ \dfrac{E_x}{c} & 0 & B_z & -B_y \\ \dfrac{E_y}{c} & -B_z & 0 & B_x \\ \dfrac{E_z}{c} & B_y & -B_x & 0 \end{pmatrix} \begin{pmatrix} 1 & 0 & 0 & 0 \\ 0 & -1 & 0 & 0 \\ 0 & 0 & -1 & 0 \\ 0 & 0 & 0 & -1 \end{pmatrix}, \tag{4.81}$$

which result in

$$[\Omega^{\alpha\beta}] = \begin{pmatrix} 0 & -\dfrac{E_x}{c} & -\dfrac{E_y}{c} & -\dfrac{E_z}{c} \\ \dfrac{E_x}{c} & 0 & -B_z & B_y \\ \dfrac{E_y}{c} & B_z & 0 & -B_x \\ \dfrac{E_z}{c} & -B_y & B_x & 0 \end{pmatrix}. \tag{4.82}$$

We use the contravariant component of the EM field tensor in equation (4.82), which we express as

$$\Omega^{\beta\kappa} = \partial^{\beta}A^{\kappa} - \partial^{\kappa}A^{\beta}, \tag{4.83}$$

to define a rank-3 tensor:

$$\partial_{\alpha}\Omega^{\beta\kappa} = \partial_{\alpha}\partial^{\beta}A^{\kappa} - \partial_{\alpha}\partial^{\kappa}A^{\beta}. \tag{4.84}$$

Multiplying equation (4.84) with the metric tensor $g_{\eta\kappa}$ for Minkowski space-time in Cartesian coordinates, we have

$$g_{\eta\kappa}\partial_{\alpha}\Omega^{\beta\kappa} = g_{\eta\kappa}(\partial_{\alpha}\partial^{\beta}A^{\kappa} - \partial_{\alpha}\partial^{\kappa}A^{\beta}), \tag{4.85}$$

so that taking into account the metric tensor elements are constant, one can write

$$\begin{aligned} \left(\partial_{\alpha}g_{\eta\kappa}\Omega^{\beta\kappa}\right) &= \partial_{\alpha}\partial^{\beta}g_{\eta\kappa}A^{\kappa} - \partial_{\alpha}g_{\eta\kappa}\partial^{\kappa}A^{\beta} \\ \Rightarrow \partial_{\alpha}\Omega^{\beta}_{\eta} &= \partial_{\alpha}\partial^{\beta}A_{\eta} - \partial_{\alpha}\partial_{\eta}A^{\beta}. \end{aligned} \tag{4.86}$$

Applying the metric tensor one more time, we have

$$\partial_\alpha g_{\mu\beta}\Omega^\beta_\eta = \partial_\alpha g_{\mu\beta}\partial^\beta A_\eta - \partial_\alpha \partial_\eta g_{\mu\beta}A^\beta, \tag{4.87}$$

so that one finds

$$\partial_\alpha \Omega_{\mu\eta} = \partial_\alpha \partial_\mu A_\eta - \partial_\alpha \partial_\eta A_\mu, \tag{4.88}$$

which we can rewrite as

$$\partial_\alpha \Omega_{\beta\kappa} = \partial_\alpha \partial_\beta A_\kappa - \partial_\alpha \partial_\kappa A_\beta. \tag{4.89}$$

Similarly, using

$$\partial_\kappa \Omega^{\kappa\beta} = \mu_0 j^\beta, \tag{4.90}$$

one can also write

$$\begin{aligned}
\partial_\kappa g_{\alpha\beta}\Omega^{\kappa\beta} &= \mu_0 g_{\alpha\beta} j^\beta \\
\Rightarrow \partial_\kappa \Omega^\kappa_\alpha &= \mu_0 j_\alpha \Rightarrow \partial_\alpha \Omega^\alpha_\kappa = \mu_0 j_\kappa.
\end{aligned} \tag{4.91}$$

Combining equations (4.86) and (4.91), we can then write

$$g^{\alpha\beta}\left(\partial_\alpha \partial_\beta A_\kappa - \partial_\alpha \partial_\kappa A_\beta\right) = \mu_0 j_\kappa. \tag{4.92}$$

Equation (4.92) is another form of equation (4.84). In the next example, we will show that equation (4.89) can be rewritten in terms of the field tensor.

Example 4.3. Starting from

$$\partial_\alpha \Omega_{\beta\kappa} = \partial_\alpha \partial_\beta A_\kappa - \partial_\alpha \partial_\kappa A_\beta. \tag{4.93}$$

(a) Show that the EM field tensor satisfies the equation

$$\partial_\kappa \Omega_{\alpha\beta} + \partial_\alpha \Omega_{\beta\kappa} + \partial_\beta \Omega_{\kappa\alpha} = 0. \tag{4.94}$$

(b) Using the antisymmetric property of the field tensor, show that equation (4.94) can be written as

$$\partial_\kappa \Omega_{\alpha\beta} + \partial_\alpha \Omega_{\beta\kappa} + \partial_\beta \Omega_{\kappa\alpha} = \partial_{[\kappa}\Omega_{\alpha\beta]} = 0, \tag{4.95}$$

Solution:
(a) Switching the indices α and β in the equation

$$\partial_\alpha \Omega_{\beta\kappa} = \partial_\alpha \partial_\beta A_\kappa - \partial_\alpha \partial_\kappa A_\beta, \tag{4.96}$$

one can write

$$\begin{aligned}
\partial_\beta \Omega_{\alpha\kappa} &= \partial_\beta \partial_\alpha A_\kappa - \partial_\beta \partial_\kappa A_\alpha \\
\Rightarrow \partial_\beta \Omega_{\kappa\alpha} &= -\left(\partial_\beta \partial_\alpha A_\kappa - \partial_\beta \partial_\kappa A_\alpha\right) = -\partial_\beta \partial_\alpha A_\kappa + \partial_\beta \partial_\kappa A_\alpha,
\end{aligned} \tag{4.97}$$

and then, switching the indices α and κ in equation (4.96),

$$\partial_\kappa \Omega_{\beta\alpha} = \partial_\kappa \partial_\beta A_\alpha - \partial_\kappa \partial_\alpha A_\beta$$
$$\Rightarrow \partial_\kappa \Omega_{\alpha\beta} = -\left(\partial_\kappa \partial_\beta A_\alpha - \partial_\kappa \partial_\alpha A_\beta\right) = -\partial_\kappa \partial_\beta A_\alpha + \partial_\kappa \partial_\alpha A_\beta. \tag{4.98}$$

Note that in equations (4.97) and (4.98), we used the antisymmetric property of the field tensor. Now, adding equations (4.96)–(4.98), we find

$$\partial_\kappa \Omega_{\alpha\beta} + \partial_\alpha \Omega_{\beta\kappa} + \partial_\beta \Omega_{\kappa\alpha}$$
$$= -\partial_\kappa \partial_\beta A_\alpha + \partial_\kappa \partial_\alpha A_\beta + \partial_\alpha \partial_\beta A_\kappa - \partial_\alpha \partial_\kappa A_\beta - \partial_\beta \partial_\alpha A_\kappa + \partial_\beta \partial_\kappa A_\alpha \tag{4.99}$$
$$\Rightarrow \partial_\kappa \Omega_{\alpha\beta} + \partial_\alpha \Omega_{\beta\kappa} + \partial_\beta \Omega_{\kappa\alpha} = 0.$$

(b) Recalling that

$$t_{[\alpha_1\alpha_2\cdots\alpha_N]} = \frac{1}{N!}(\text{Alternating subtraction and addition over} \tag{4.100}$$
$$\text{all permutations of the indices } \alpha_1\alpha_2\cdots\alpha_N),$$

for a rank-3 tensor, we have

$$t_{[\alpha_1\alpha_2\alpha_3]} = \frac{1}{3!}(t_{\alpha_1\alpha_2\alpha_3} - t_{\alpha_1\alpha_3\alpha_2} + t_{\alpha_3\alpha_1\alpha_2} - t_{\alpha_3\alpha_2\alpha_1} + t_{\alpha_2\alpha_3\alpha_1} - t_{\alpha_2\alpha_1\alpha_3}), \tag{4.101}$$

so that one can write

$$\partial_{[\kappa}\Omega_{\alpha\beta]} = \frac{1}{3!}(\partial_\kappa \Omega_{\alpha\beta} - \partial_\kappa \Omega_{\beta\alpha} + \partial_\beta \Omega_{\kappa\alpha} - \partial_\beta \Omega_{\alpha\kappa} + \partial_\alpha \Omega_{\beta\kappa} - \partial_\alpha \Omega_{\kappa\beta}). \tag{4.102}$$

Noting that the EM field tensor is antisymmetric, one can establish the relations

$$\partial_\kappa \Omega_{\beta\alpha} = -\partial_\kappa \Omega_{\alpha\beta}, \; \partial_\beta \Omega_{\alpha\kappa} = -\partial_\beta \Omega_{\kappa\alpha}, \; \partial_\alpha \Omega_{\kappa\beta} = -\partial_\alpha \Omega_{\beta\kappa}. \tag{4.103}$$

Upon substituting equation (4.103) into equation (4.102), we find

$$\partial_{[\kappa}\Omega_{\alpha\beta]} = \frac{2}{3!}(\partial_\kappa \Omega_{\alpha\beta} + \partial_\alpha \Omega_{\beta\kappa} + \partial_\beta \Omega_{\kappa\alpha})$$
$$\Rightarrow \partial_\kappa \Omega_{\alpha\beta} + \partial_\alpha \Omega_{\beta\kappa} + \partial_\beta \Omega_{\kappa\alpha} = 0 \Rightarrow \frac{3!}{2}\partial_{[\kappa}\Omega_{\alpha\beta]} = 0 \tag{4.104}$$
$$\Rightarrow \partial_{[\kappa}\Omega_{\alpha\beta]} = 0.$$

In view of the result in example 4.1, the complete EM field equations can be written as

$$\partial_{[\kappa}\Omega_{\alpha\beta]} = 0, \; \partial_\alpha \Omega^{\alpha\beta} = \mu_0 j^\beta, \tag{4.105}$$

which are very simple equations.

4.5 Gauge transformation

In electromagnetism, it is known that we can choose the three-vector potential $\vec{A}$ without changing the electric or magnetic fields:

$$\vec{E} = -\frac{\partial \vec{A}}{\partial t} - \nabla V, \; \vec{B} = \nabla \times \vec{A}. \tag{4.106}$$

This is known as gauge transformation. There are two gauge transformations: Coulomb gauge and Lorentz gauge.

Coulomb gauge

In a Coulomb gauge we can choose a vector potential, $\vec{A}(\vec{r}, t)$, such that

$$\nabla \cdot \vec{A}(\vec{r}, t) = 0, \tag{4.107}$$

without changing the electric and magnetic fields. Noting that

$$\nabla \cdot \vec{A}(\vec{r}, t) = \frac{\partial A^1}{\partial x^1} + \frac{\partial A^2}{\partial x^2} + \frac{\partial A^3}{\partial x^3} = 0, \tag{4.108}$$

using the four-vector potential and the four-gradient, one finds

$$\begin{aligned} \partial_\alpha A^\alpha &= \frac{\partial A^0}{\partial x^0} + \frac{\partial A^1}{\partial x^1} + \frac{\partial A^2}{\partial x^2} + \frac{\partial A^3}{\partial x^3} \\ &\Rightarrow \nabla \cdot \vec{A}(\vec{r}, t) = \partial_\alpha A^\alpha - \partial_0 A^0 = 0 \\ &\Rightarrow \partial_\alpha A^\alpha = \partial_0 A^0. \end{aligned} \tag{4.109}$$

Equation (4.109) is a Coulomb gauge in Minkowski space-time.

Lorentz gauge

In a Lorentz gauge we choose the vector potential such that

$$\nabla \cdot \vec{A} = -\frac{1}{c^2}\frac{\partial V}{\partial t}. \tag{4.110}$$

Similarly, this can be rewritten as

$$\begin{aligned} \frac{\partial A^1}{\partial x^1} + \frac{\partial A^2}{\partial x^2} + \frac{\partial A^3}{\partial x^3} &= -\frac{\partial}{\partial ct}\left(\frac{V}{c}\right) = -\frac{\partial A^0}{\partial x^0} \\ &\Rightarrow \frac{\partial A^0}{\partial x^0} + \frac{\partial A^1}{\partial x^1} + \frac{\partial A^2}{\partial x^2} + \frac{\partial A^3}{\partial x^3} = 0 \\ &\Rightarrow \partial_\alpha A^\alpha = 0. \end{aligned} \tag{4.111}$$

Equation (4.111) is a Lorentz gauge in Minkowski space-time.

4.6 Maxwell's equations in a Lorentz gauge

In this section we will re-derive Maxwell's equation in a Lorentz gauge from the field equations. To this end, we note that

$$\partial_\beta A^\beta = 0 \Rightarrow \partial_\alpha \partial_\beta A^\beta = 0 \Rightarrow \partial_\alpha \partial_\beta g^{\beta\kappa} A_\kappa = 0$$
$$\Rightarrow \partial_\alpha g^{\beta\kappa} \partial_\beta A_\kappa = 0 \Rightarrow \partial_\alpha \partial^\kappa A_\kappa = 0,$$

(4.112)

where we used

$$\partial^\kappa = g^{\beta\kappa} \partial_\beta .$$

(4.113)

The field equation

$$g^{\alpha\beta}\left(\partial_\alpha \partial_\beta A_\kappa - \partial_\alpha \partial_\kappa A_\beta\right) = \mu_0 j_\kappa ,$$

(4.114)

can then be rewrite as

$$\left(\partial_\alpha g^{\alpha\beta} \partial_\beta A_\kappa - g^{\alpha\beta} \partial_\alpha \partial_\kappa A_\beta\right) = \mu_0 j_\kappa \Rightarrow \left(\partial_\alpha \partial^\alpha A_\kappa - \partial^\beta \partial_\kappa A_\beta\right) = \mu_0 j_\kappa$$
$$\Rightarrow \left(\partial_\alpha \partial^\alpha A_\kappa - \partial_\kappa \partial^\beta A_\beta\right) = \mu_0 j_\kappa .$$

(4.115)

Since in the Lorentz gauge

$$\partial_\alpha \partial^\beta A_\beta = 0,$$

(4.116)

the field equation becomes

$$\partial_\alpha \partial^\alpha A_\kappa = \mu_0 j_\kappa .$$

(4.117)

Using the d'Alembert operator, one can put the field equation in the form

$$\partial_\alpha \partial^\alpha A_\kappa = \mu_0 j_\kappa \Rightarrow \Box A_\kappa = \mu_0 j_\kappa .$$

(4.118)

Multiplying this with the contravariant component of the metric tensor, we find

$$\Box g^{\alpha\kappa} A_\kappa = \mu_0 g^{\alpha\kappa} j_\kappa \Rightarrow \Box A^\alpha = \mu_0 j^\alpha .$$

(4.119)

Recalling that

$$[A^\beta] = \left(\frac{V}{c}, \vec{A}\right) = \left(\frac{V}{c}, A_x, A_y, A_z\right),$$

(4.120)

in the Lorentz gauge

$$\partial_\alpha A^\alpha = 0 \Rightarrow \left(\frac{\partial}{c\partial t}\left(\frac{V}{c}\right), \nabla \cdot \vec{A}\right) = 0 \Rightarrow \frac{1}{c^2}\frac{\partial V}{\partial t} + \nabla \cdot \vec{A} = 0$$
$$\Rightarrow \nabla \cdot \vec{A} = -\frac{1}{c^2}\frac{\partial V}{\partial t}.$$

(4.121)

From the four-field equations,

$$\partial_\alpha \partial^\alpha A^\kappa = \Box A^\kappa = \frac{1}{c^2}\frac{\partial^2 A^\kappa}{\partial t^2} - \frac{\partial^2 A^\kappa}{\partial x^2} - \frac{\partial^2 A^\kappa}{\partial y^2} - \frac{\partial^2 A^\kappa}{\partial z^2} = \mu_0 j^\kappa ,$$

(4.122)

we find

$$\frac{1}{c^2}\frac{\partial^2 A^0}{\partial t^2} - \frac{\partial^2 A^0}{\partial x^2} - \frac{\partial^2 A^0}{\partial y^2} - \frac{\partial^2 A^0}{\partial z^2} = \mu_0 j^0,$$ (4.123)

$$\frac{1}{c^2}\frac{\partial^2 A^1}{\partial t^2} - \frac{\partial^2 A^1}{\partial x^2} - \frac{\partial^2 A^1}{\partial y^2} - \frac{\partial^2 A^1}{\partial z^2} = \mu_0 j^1,$$ (4.124)

$$\frac{1}{c^2}\frac{\partial^2 A^2}{\partial t^2} - \frac{\partial^2 A^2}{\partial x^2} - \frac{\partial^2 A^2}{\partial y^2} - \frac{\partial^2 A^2}{\partial z^2} = \mu_0 j^2,$$ (4.125)

$$\frac{1}{c^2}\frac{\partial^2 A^3}{\partial t^2} - \frac{\partial^2 A^3}{\partial x^2} - \frac{\partial^2 A^3}{\partial y^2} - \frac{\partial^2 A^3}{\partial z^2} = \mu_0 j^3.$$ (4.126)

First let us consider equation (4.123). Using the component $j^0 = c\rho$ from the four-current density,

$$[j^\alpha] = (c\rho, \vec{j}\,),$$ (4.127)

equation (4.123) can be put in the form

$$\frac{1}{c^2}\frac{\partial^2}{\partial t^2}\left(\frac{V}{c}\right) - \frac{\partial^2}{\partial x^2}\left(\frac{V}{c}\right) - \frac{\partial^2}{\partial y^2}\left(\frac{V}{c}\right) - \frac{\partial^2}{\partial z^2}\left(\frac{V}{c}\right) = c\mu_0\rho$$

$$\frac{1}{c^2}\frac{\partial^2 V}{\partial t^2} - \nabla^2 V = c^2\mu_0\rho = \frac{\rho}{\epsilon_0},$$ (4.128)

where we used the speed of light in vacuum:

$$c = \frac{1}{\sqrt{\epsilon_0\mu_0}}.$$ (4.129)

Introducing a vector field, $\vec{E}$ (electric field vector), defined by

$$\vec{E} = -\frac{\partial \vec{A}}{\partial t} - \nabla V \Rightarrow \nabla \cdot \vec{E} = -\left[\frac{\partial}{\partial t}(\nabla \cdot \vec{A}) + \nabla^2 V\right],$$ (4.130)

in the Lorentz gauge, where

$$\nabla \cdot \vec{A} = -\frac{1}{c^2}\frac{\partial V}{\partial t},$$ (4.131)

one can write

$$\nabla \cdot \vec{E} = \frac{1}{c^2}\frac{\partial^2 V}{\partial t^2} - \nabla^2 V.$$ (4.132)

Then substituting equation (4.132) into equation (4.128), we find

$$\nabla \cdot \vec{E} = \frac{\rho}{\epsilon_0}, \quad \text{(Gauss's law)}.$$

Adding equations (4.124)–(4.126) and noting that the three-vector potential and the three-current densities

$$\vec{A} = (A^1, A^2, A^3), \vec{j} = (j^1, j^2, j^3),$$

one finds

$$\frac{1}{c^2}\frac{\partial^2 \vec{A}}{\partial t^2} - \nabla^2 \vec{A} = \mu_0 \vec{j}. \qquad (4.133)$$

Introducing a second vector field, $\vec{B}$ (the magnetic field), defined by

$$\vec{B} = \nabla \times \vec{A}, \qquad (4.134)$$

we have

$$\nabla \times \vec{B} = \nabla \times (\nabla \times \vec{A}) = \nabla(\nabla \cdot \vec{A}) - \nabla^2 \vec{A}, \qquad (4.135)$$

so that in the Lorentz gauge, using equation (4.121), one can write

$$\nabla \times \vec{B} = \nabla\left(-\frac{1}{c^2}\frac{\partial V}{\partial t}\right) - \nabla^2 \vec{A} = -\frac{1}{c^2}\frac{\partial(\nabla V)}{\partial t} - \nabla^2 \vec{A}$$

$$\Rightarrow \nabla^2 \vec{A} = -\nabla \times \vec{B} - \frac{1}{c^2}\frac{\partial(\nabla V)}{\partial t}. \qquad (4.136)$$

Substituting equation (4.136) into equation (4.133), one finds

$$\nabla \times \vec{B} + \frac{1}{c^2}\frac{\partial^2 \vec{A}}{\partial t^2} + \frac{1}{c^2}\frac{\partial(\nabla V)}{\partial t} = \nabla \times \vec{B} + \frac{1}{c^2}\frac{\partial}{\partial t}\left[\nabla V + \frac{\partial \vec{A}}{\partial t}\right] = \mu_0 \vec{j}, \qquad (4.137)$$

and recalling that

$$\vec{E} = -\nabla V - \frac{\partial \vec{A}}{\partial t}, \qquad (4.138)$$

one finds

$$\nabla \times \vec{B} - \frac{1}{c^2}\frac{\partial \vec{E}}{\partial t} = \mu_0 \vec{j}$$

$$\Rightarrow \nabla \times \vec{B} = \epsilon_0 \mu_0 \frac{\partial \vec{E}}{\partial t} + \mu_0 \vec{j} \text{ (Ampere's law).} \qquad (4.139)$$

The other two Maxwell's equations can easily be obtained using properties of vector calculus:

$$\nabla \times \vec{E} = \nabla \times \left(-\nabla V - \frac{\partial \vec{A}}{\partial t}\right) = \left(-\nabla \times \nabla V - \frac{\partial}{\partial t}(\nabla \times \vec{A})\right)$$

$$= -\nabla \times (\nabla V) - \frac{\partial \vec{B}}{\partial t} \Rightarrow \nabla \times \vec{E} = -\frac{\partial \vec{B}}{\partial t} \text{ (Faraday's law),} \qquad (4.140)$$

where we used the relation

$$\nabla \times (\nabla V) = 0. \tag{4.141}$$

For the magnetic field,

$$\nabla \cdot \vec{B} = \nabla \cdot \left(\nabla \times \vec{A}\right), \tag{4.142}$$

since for any vector $\vec{A}$,

$$\nabla \cdot \left(\nabla \times \vec{A}\right) = 0, \tag{4.143}$$

we can easily see that

$$\nabla \cdot \vec{B} = 0 \text{ (No name law!).} \tag{4.144}$$

In another inertial reference frame (i.e., the S' frame), the EM field tensor can easily be determined from the vector potential or the EM field tensor in the S frame using

$$A'^{\alpha} = \Lambda^{\alpha}_{\beta} A^{\beta}, \; \Omega'^{\alpha\beta} = \Lambda^{\alpha}_{\kappa} \Lambda^{\beta}_{\mu} \Omega^{\kappa\mu}. \tag{4.145}$$

4.7 Charged particle equation of motion

Consider a particle of rest mass m_0 in a region where there is an EM field. We recall that EM force is a pure force and it does not alter the rest mass of the particle. Thus the equation of motion for the particle can be written as

$$F^{\alpha} = \frac{dp^{\alpha}}{d\tau} = m_0 \frac{du^{\alpha}}{d\tau}. \tag{4.146}$$

We recall the force in terms of the field tensor

$$F_{\alpha} = q\Omega_{\alpha\beta} u^{\beta} \Rightarrow F^{\beta} = g^{\beta\alpha} F_{\alpha} = qg^{\beta\alpha}\Omega_{\alpha\kappa}u^{\kappa} = q\Omega^{\beta}_{\kappa}u^{\kappa}, \tag{4.147}$$

and the equation of motion becomes

$$m_0 \frac{du^{\alpha}}{d\tau} = q\Omega^{\alpha}_{\beta} u^{\beta}. \tag{4.148}$$

This equation of motion is in Cartesian coordinates. We recall, generally for any coordinates, that the intrinsic derivative for the contravariant component is given by

$$\frac{Du^{\alpha}}{D\tau} = \frac{du^{\alpha}}{d\tau} + \Gamma^{\alpha}_{\kappa\beta} u^{\kappa} \frac{dx^{\beta}}{d\tau}, \tag{4.149}$$

and the covariant component by

$$\frac{Du_{\alpha}}{D\tau} = \frac{du_{\alpha}}{d\tau} - \Gamma^{\beta}_{\alpha\kappa} u_{\beta} \frac{dx^{\kappa}}{d\tau}. \tag{4.150}$$

For an arbitrary set of coordinates, one can then write the contravariant component of the force in terms of the corresponding components for the momentum:

$$F^\alpha = \frac{Dp^\alpha}{D\tau} = m_0 \frac{Du^\alpha}{D\tau} = m_0 \left(\frac{du^\alpha}{d\tau} + \Gamma^\alpha_{\kappa\beta} u^\kappa \frac{dx^\beta}{d\tau} \right), \tag{4.151}$$

where still *the EM force is pure*. Then the equation of motion in terms of the contravariant component can be written as

$$m_0 \left(\frac{du^\alpha}{d\tau} + \Gamma^\alpha_{\kappa\beta} u^\kappa \frac{dx^\beta}{d\tau} \right) = q\Omega^\alpha_\beta u^\beta. \tag{4.152}$$

In terms of only the four-velocity,

$$\frac{dx^\beta}{d\tau} = u^\beta, \tag{4.153}$$

we may write

$$m_0 \left(\frac{du^\alpha}{d\tau} + \Gamma^\alpha_{\kappa\beta} u^\beta u^\kappa \right) = q\Omega^\alpha_\beta u^\beta. \tag{4.154}$$

Or in terms of only the coordinates,

$$m_0 \left(\frac{d^2 x^\alpha}{d\tau^2} + \Gamma^\alpha_{\beta\kappa} \frac{dx^\beta}{d\tau} \frac{dx^\kappa}{d\tau} \right) = q\Omega^\alpha_\beta \frac{dx^\beta}{d\tau}, \tag{4.155}$$

which we can put in the form

$$\frac{d^2 x^\alpha}{d\tau^2} + \Gamma^\alpha_{\beta\kappa} \frac{dx^\beta}{d\tau} \frac{dx^\kappa}{d\tau} = \frac{q}{m_0} \Omega^\alpha_\beta \frac{dx^\beta}{d\tau}. \tag{4.156}$$

In the absence of an EM field or for a non-charged particle (i.e. $\Omega^\alpha_\beta = 0$ or $q = 0$), we find

$$\frac{d^2 x^\alpha}{d\tau^2} + \Gamma^\alpha_{\beta\kappa} \frac{dx^\beta}{d\tau} \frac{dx^\kappa}{d\tau} = 0, \tag{4.157}$$

which means the motion of the particle is *geodesic*.

4.8 Homework assignment

Problem 1. For an observer in the S' frame traveling with a velocity $\vec{v}$ along the positive x-axis as measured by an observer in the S frame, using Lorentz transformation for $x'^\alpha \to x^\alpha$,

$$x^0 = \gamma x'^0 + \gamma\beta x'^1, \ x^1 = \gamma\beta x'^0 + \gamma x'^1, \ x^2 = x'^2, \ x^3 = x'^3,$$

or $x^\alpha \to x'^\alpha$,

$$x'^0 = \gamma x^0 - \gamma \beta x^1, \ x'^1 = \gamma x^1 - \gamma \beta x^0, \ x'^2 = x^2, \ x'^3 = x^3,$$

where

$$\beta = \frac{v}{c}, \ \gamma = \frac{1}{\sqrt{1 - \dfrac{v^2}{c^2}}},$$

which gives the transformation matrices

$$[\Lambda^\kappa{}_\alpha] = \left[\frac{\partial x^\kappa}{\partial x'^\alpha}\right], \ [\Lambda'^\beta{}_\mu] = \left[\frac{\partial x'^\beta}{\partial x^\mu}\right].$$

Find the EM field tensor $\overleftrightarrow{\Omega}'$, covariant, contravariant, and mixed components.

Problem 2. Starting from

$$\vec{E} = -\frac{\partial \vec{A}}{\partial t} - \nabla V, \ \vec{B} = \nabla \times \vec{A},$$

show that the contravariant and covariant components of the EM field tensor are given by

$$[\Omega_{\alpha\beta}] = \begin{pmatrix} 0 & \dfrac{E_x}{c} & \dfrac{E_y}{c} & \dfrac{E_z}{c} \\ -\dfrac{E_x}{c} & 0 & -B_z & B_y \\ -\dfrac{E_y}{c} & B_z & 0 & -B_x \\ -\dfrac{E_z}{c} & -B_y & B_x & 0 \end{pmatrix},$$

and

$$[\Omega^{\alpha\beta}] = \begin{pmatrix} 0 & -\dfrac{E_x}{c} & -\dfrac{E_y}{c} & -\dfrac{E_z}{c} \\ \dfrac{E_x}{c} & 0 & -B_z & B_y \\ \dfrac{E_y}{c} & B_z & 0 & -B_x \\ \dfrac{E_z}{c} & -B_y & B_x & 0 \end{pmatrix}.$$

Problem 3. Starting from the EM field equations,

$$\partial_{[\kappa}\Omega_{\alpha\beta]} = 0, \ \partial_\alpha\Omega^{\alpha\beta} = \mu_0 j^\beta,$$

derive the Maxwell's equations.

Problem 4. Show that $\vec{E} \cdot \vec{B}$ is Lorentz invariant.

Problem 5. In the S frame, show that the three-acceleration of a charged particle q in an EM field is given by

$$\vec{a} = \frac{d\vec{u}}{dt} = \frac{q}{\gamma_u m_0}\left[\vec{E} + \vec{u} \times \vec{B} - \frac{1}{c^2}\left(\vec{u} \cdot \vec{E}\right)\vec{u}\right].$$

References

[1] Hobson M P, Efstathiou G P and Lasenby A N 2006 *General Relativity-An Introduction for Physicists* (Cambridge: Univ. Cambridge Press)

[2] Erenso D and Montemayor V 2022 *Studies in Theoretical Physics Volume 1: Fundamental mathematical methods* (Bristol: IOP Publishing)

[3] Arfken G B and Weber H-J 2005 *Mathematical Methods for Physicists* 6th ed. (Boston, MA: Elsevier)

[4] Arfken G B 1985 *Mathematical Methods for Physicists* 3rd ed. (Orlando, FL: Academic Press)

[5] Boas M L 2006 *Mathematical Methods in the Physical Sciences* 3rd ed. (New York: John Wiley & Sons)

Printed in the USA
CPSIA information can be obtained
at www.ICGtesting.com
CBHW082128290324
5693CB00021B/279